Impacts of Climate Change on Tree Physiology and Responses of Forest Ecosystems

Impacts of Climate Change on Tree Physiology and Responses of Forest Ecosystems

Editor

Mariangela Fotelli

MDPI • Basel • Beijing • Wuhan • Barcelona • Belgrade • Manchester • Tokyo • Cluj • Tianjin

Editor
Mariangela Fotelli
Hellenic Agricultural Organization
Dimitra Forest Research Institute
Vassilika, Thessaloniki
Greece

Editorial Office
MDPI
St. Alban-Anlage 66
4052 Basel, Switzerland

This is a reprint of articles from the Special Issue published online in the open access journal *Forests* (ISSN 1999-4907) (available at: https://www.mdpi.com/journal/forests/special_issues/Tree_Physiology).

For citation purposes, cite each article independently as indicated on the article page online and as indicated below:

LastName, A.A.; LastName, B.B.; LastName, C.C. Article Title. *Journal Name* **Year**, *Volume Number*, Page Range.

ISBN 978-3-0365-2750-5 (Hbk)
ISBN 978-3-0365-2751-2 (PDF)

Cover image courtesy of Mariangela Fotelli

Contents

About the Editor . . . . **vii**

Mariangela N. Fotelli
Impacts of Climate Change on Tree Physiology and Responses of Forest Ecosystems
Reprinted from: *Forests* **2021**, *12*, 1728, doi:10.3390/f12121728 . . . . **1**

Tae-Lim Kim, Hoyong Chung, Karpagam Veerappan, Wi Young Lee, Danbe Park and Hyemin Lim
Physiological and Transcriptome Responses to Elevated CO_2 Concentration in *Populus*
Reprinted from: *Forests* **2021**, *12*, 980, doi:10.3390/f12080980 . . . . **5**

Filippos Bantis, Julia Graap, Elena Früchtenicht, Filippo Bussotti, Kalliopi Radoglou and Wolfgang Brüggemann
Field Performances of Mediterranean Oaks in Replicate Common Gardens for Future Reforestation under Climate Change in Central and Southern Europe: First Results from a Four-Year Study
Reprinted from: *Forests* **2021**, *12*, 678, doi:10.3390/f12060678 . . . . **23**

Romà Ogaya and Josep Peñuelas
Climate Change Effects in a Mediterranean Forest Following 21 Consecutive Years of Experimental Drought
Reprinted from: *Forests* **2021**, *12*, 306, doi:10.3390/f12030306 . . . . **39**

Evangelia Korakaki and Mariangela N. Fotelli
Sap Flow in Aleppo Pine in Greece in Relation to Sapwood Radial Gradient, Temporal and Climatic Variability
Reprinted from: *Forests* **2021**, *12*, 2, doi:10.3390/f12010002 . . . . **51**

Peter Petrík, Anja Petek, Alena Konôpková, Michal Bosela, Peter Fleischer, Josef Frýdl and Daniel Kurjak
Stomatal and Leaf Morphology Response of European Beech (*Fagus sylvatica* L.) Provenances Transferred to Contrasting Climatic Conditions
Reprinted from: *Forests* **2020**, *11*, 1359, doi:10.3390/f11121359 . . . . **67**

Athanasios Zindros, Kalliopi Radoglou, Elias Milios and Kyriaki Kitikidou
Tree Line Shift in the Olympus Mountain (Greece) and Climate Change
Reprinted from: *Forests* **2020**, *11*, 985, doi:10.3390/f11090985 . . . . **87**

Xiao Wang, Xiaoli Wei, Gaoyin Wu and Shenqun Chen
High Nitrate or Ammonium Applications Alleviated Photosynthetic Decline of *Phoebe bournei* Seedlings under Elevated Carbon Dioxide
Reprinted from: *Forests* **2020**, *11*, 293, doi:10.3390/f11120293 . . . . **95**

Joanna Kijowska-Oberc, Aleksandra M. Staszak, Jan Kamiński and Ewelina Ratajczak
Adaptation of Forest Trees to Rapidly Changing Climate
Reprinted from: *Forests* **2020**, *11*, 123, doi:10.3390/f11020123 . . . . **109**

Paula E. Marquardt, Brian R. Miranda and Frank W. Telewski
Shifts in Climate–Growth Relationships of Sky Island Pines
Reprinted from: *Forests* **2019**, *10*, 1011, doi:10.3390/f10111011 . . . . **133**

Salvador Sampayo-Maldonado, Cesar A. Ordoñez-Salanueva, Efisio Mattana, Tiziana Ulian,
Michael Way, Elena Castillo-Lorenzo, Patricia D. Dávila-Aranda, Rafael Lira-Saade, Oswaldo
Téllez-Valdéz, Norma I. Rodriguez-Arevalo and Cesar M. Flores-Ortíz
Thermal Time and Cardinal Temperatures for Germination of *Cedrela odorata* L.
Reprinted from: *Forests* **2019**, *10*, 841, doi:10.3390/f10100841 . . . . **145**

About the Editor

Mariangela Fotelli graduated with distinction from the Aristotle University of Thessaloniki, Greece, and received her PhD with honors in 2002 from the Albert Ludwig University of Freiburg, Germany, working on nitrogen and water balance in European beech natural regeneration. In the following years, she participated in several international research projects and carried out post-doctoral studies on the carbon metabolism of N-fixing legumes at the Agricultural University of Athens and on the nitrogen ecophysiology, water and carbon status of forest species under climatic variability at the Democritus University of Thrace and the Forest Research Institute of Thessaloniki, Greece. She has been a researcher and head of the lab of Forest Ecophysiology at the Forest Research Institute of the Hellenic Agricultural Organization Dimitra since 2017.

Editorial

Impacts of Climate Change on Tree Physiology and Responses of Forest Ecosystems

Mariangela N. Fotelli

Hellenic Agricultural Organization Dimitra, Forest Research Institute, Vassilika, 57006 Thessaloniki, Greece;
fotelli@fri.gr

Citation: Fotelli, M.N. Impacts of Climate Change on Tree Physiology and Responses of Forest Ecosystems. *Forests* **2021**, *12*, 1728. https://doi.org/10.3390/f12121728

Received: 5 December 2021
Accepted: 6 December 2021
Published: 8 December 2021

Publisher's Note: MDPI stays neutral with regard to jurisdictional claims in published maps and institutional affiliations.

1. Tree Adaptive Responses to Climatic Variability

In a changing climate, forest trees have to deal with a range of altered environmental conditions. Within this Special Issue, researchers aimed at revealing the responses of different forest species to various globally changing abiotic parameters, such as increasing atmospheric CO_2, elevated temperatures, and limited water availability.

In an experiment with open-top chambers, Kim et al. [1] studied the physiological and transcriptome responses of Poplar hybrids to elevated CO_2, being one of the main components of global change. They identified differentially expressed genes and alterations in chlorophylls, starch, and soluble sugars in response to increasing CO_2. The impact of elevated CO_2 was also assessed by Wang et al. [2], who reported photosynthetic declines in *Phoebe bournei* Hemsl., an endemic tree threatened by habitat loss in China. However, they concluded that this negative impact of increased CO_2 may be alleviated by the application of ammonium or nitrate fertilization. The effect of drier growth conditions on the tree radial profile of sap flow was assessed in a coastal Mediterranean Aleppo pine ecosystem [3]. A steeper decline in sap flux density with increasing sapwood depth was observed in drier periods. The limiting role of water availability in such xerothermic environments was further supported by the fact that sap flow responded positively to a favorable microclimate only when water availability exceeded a certain threshold. The effect of rising temperature was studied by Sambayo-Maldonado et al. [4] who assessed the thermal time and cardinal temperatures of *Cedrela odorata* in order to estimate changes in the germination of red cedar under different climate change scenarios. It was found that as the temperature increases in line with the predictions for 2050 and 2100, the germination of the soil seed bank will accelerate, which may have implications for the future distribution of this important forest species in tropical environments.

Common gardens and provenance trials are an experimental approach offering long-term information regarding the potential of different forest species and origins to adapt to future climate. In order to assess the suitability of different Mediterranean oak species for potential assisted migration schemes in central Europe, Bantis et al. [5], established a four common garden experiment and evaluated the early field performance of oak seedlings of different species and provenances. Based on survival rates, morphological, and physiological features, they concluded that *Quercus pubescens* may be a suitable candidate forest tree for poor and dry soils under the expected summer heat and drought conditions in Central Europe. In the same context, a provenance trials approach was used to evaluate nine European beech (*Fagus sylvatica* L.) provenances under contrasting climates [6]. An acclimation potential of provenances transferred to warmer and drier conditions was indicated by the development of smaller leaves with decreasing stomatal density and length of guard cells. However, the higher phenotypic plasticity in stomatal traits was also associated with increasing mortality under more arid conditions.

2. Responses to Changing Climate at the Ecosystem Level

Climate change is also expected to induce alternations in the distribution range of forest species, species' replacement and losses of forest land, with important economic and ecological consequences.

In this context, Zindros et al. [7] monitored the changes in tree line over a 40-year period in the core of Olympus Mt. National Park in Greece and reported a warming-related upward shift in the tree line. Apart from spatial shifts, temporal shifts in forest ecosystems ecophysiological and phenological responses are also related to climate change. Marquardt et al. [8], compared the sensitivity and temporal stability of climate–growth relationships in *Pinus arizonica*, and *P. ponderosa* var. *brachyptera* consisting of two needle-types, in Arizona, USA, in order to characterize their growth responses to changing climate. They concluded that the species responded to summer precipitation pre-1950 but are now responding to spring precipitation post-1950, representing a shift in allocation of tree resources from maximizing biomass in the summer to reproduction spring. In the same lines, Ogaya and Peñuelas [9] found that a small but continuous decrease in soil water availability for 21 years in a Mediterranean holm oak (*Quercus ilex* L.) forest resulted in the gradual replacement of holm oak by another evergreen shrub (*Phyllirea latifolia* L.), which is more tolerant to hot and dry conditions.

3. Conclusions

The above-mentioned and other forest trees' and ecosystem responses to rapidly changing climate are reviewed by Kijowska-Oberc et al. [10]. In their review, they brought together literature reports on a range of adaptive responses to both forest abiotic threats (such as drought and heat) and biotic threats (such as pathogens and insects) and they discuss the role of epigenetics, phenology, and phenotypic plasticity for tree adaptations to climate change.

The studies included in this Special Issue illustrate the clear effects of climate change on forest trees' function, evidenced by different responses on genes expression, biochemical, and physiological levels. However, it is also indicated that different forest species from a broad range of climate zones and ecosystem types are able to develop adaptive mechanisms to cope with the current climatic variability. On the ecosystem level this may lead to distribution shifts and changes in vegetation composition. Further research is needed to identify if measures, such as assisted migration or others, may facilitate the transition to climate change resilient forest ecosystems.

Funding: This research received no external funding.

Data Availability Statement: Not applicable.

Conflicts of Interest: The author declares no conflict of interest.

References

1. Kim, T.-L.; Chung, H.; Veerappan, K.; Lee, W.Y.; Park, D.; Lim, H. Physiological and Transcriptome Responses to Elevated CO_2 Concentration in *Populus*. *Forests* **2021**, *12*, 980. [CrossRef]
2. Wang, X.; Wei, X.; Wu, G.; Chen, S. High Nitrate or Ammonium Applications Alleviated Photosynthetic Decline of *Phoebe bournei* Seedlings under Elevated Carbon Dioxide. *Forests* **2020**, *11*, 293. [CrossRef]
3. Korakaki, E.; Fotelli, M.N. Sap Flow in Aleppo Pine in Greece in Relation to Sapwood Radial Gradient, Temporal and Climatic Variability. *Forests* **2021**, *12*, 2. [CrossRef]
4. Sampayo-Maldonado, S.; Ordoñez-Salanueva, C.A.; Mattana, E.; Ulian, T.; Way, M.; Castillo-Lorenzo, E.; Dávila-Aranda, P.D.; Lira-Saade, R.; Téllez-Valdéz, O.; Rodriguez-Arevalo, N.I.; et al. Thermal Time and Cardinal Temperatures for Germination of *Cedrela odorata* L. *Forests* **2019**, *10*, 841. [CrossRef]
5. Bantis, F.; Graap, J.; Früchtenicht, E.; Bussotti, F.; Radoglou, K.; Brüggemann, W. Field Performances of Mediterranean Oaks in Replicate Common Gardens for Future Reforestation under Climate Change in Central and Southern Europe: First Results from a Four-Year Study. *Forests* **2021**, *12*, 678. [CrossRef]
6. Petrík, P.; Petek, A.; Konôpková, A.; Bosela, M.; Fleischer, P.; Frýdl, J.; Kurjak, D. Stomatal and Leaf Morphology Response of European Beech (*Fagus sylvatica* L.) Provenances Transferred to Contrasting Climatic Conditions. *Forests* **2020**, *11*, 1359. [CrossRef]

7. Zindros, A.; Radoglou, K.; Milios, E.; Kitikidou, K. Tree Line Shift in the Olympus Mountain (Greece) and Climate Change. *Forests* **2020**, *11*, 985. [CrossRef]
8. Marquardt, P.E.; Miranda, B.R.; Telewski, F.W. Shifts in Climate–Growth Relationships of Sky Island Pines. *Forests* **2019**, *10*, 1011. [CrossRef]
9. Ogaya, R.; Peñuelas, J. Climate Change Effects in a Mediterranean Forest Following 21 Consecutive Years of Experimental Drought. *Forests* **2021**, *12*, 306. [CrossRef]
10. Kijowska-Oberc, J.; Staszak, A.M.; Kamiński, J.; Ratajczak, E. Adaptation of Forest Trees to Rapidly Changing Climate. *Forests* **2020**, *11*, 123. [CrossRef]

forests

Article

Physiological and Transcriptome Responses to Elevated CO_2 Concentration in *Populus*

Tae-Lim Kim [1], Hoyong Chung [2], Karpagam Veerappan [2], Wi Young Lee [1], Danbe Park [1] and Hyemin Lim [1,*]

[1] Forest Bioresources Department, National Institute of Forest Science, Suwon-si 16631, Gyeonggi-do, Korea; ktlmi01@korea.kr (T.-L.K.); lwy20@korea.kr (W.Y.L.); danbepark@korea.kr (D.P.)

[2] 3BIGS CO. LTD., 156 Gwanggyo-ro, Yeongtong-gu, Suwon-si 16506, Gyeonggi-do, Korea; hychung@3bigs.com (H.C.); karpagam@3bigs.com (K.V.)

* Correspondence: supia1125@korea.kr; Tel.: +82-031-290-1116

Abstract: Global climate change is heavily affected by an increase in CO_2. As one of several efforts to cope with this, research on poplar, a representative, fast growing, and model organism in plants, is actively underway. The effects of elevated atmospheric CO_2 on the metabolism, growth, and transcriptome of poplar were investigated to predict productivity in an environment where CO_2 concentrations are increasing. Poplar trees were grown at ambient (400 ppm) or elevated CO_2 concentrations (1.4× ambient, 560 ppm, and 1.8× ambient, 720 ppm) for 16 weeks in open-top chambers (OTCs). We analyzed the differences in the transcriptomes of *Populus alba* × *Populus glandulosa* clone "Clivus" and *Populus euramericana* clone "I-476" using high-throughput sequencing techniques and elucidated the functions of the differentially expressed genes (DEGs) using various functional annotation methods. About 272,355 contigs and 207,063 unigenes were obtained from transcriptome assembly with the Trinity assembly package. Common DEGs were identified which were consistently regulated in both the elevated CO_2 concentrations. In Clivus 29, common DEGs were found, and most of these correspond to cell wall proteins, especially hydroxyproline-rich glycoproteins (HRGP), or related to fatty acid metabolism. Concomitantly, in I-476, 25 were identified, and they were related to heat shock protein (HSP) chaperone family, photosynthesis, nitrogen metabolism, and carbon metabolism. In addition, carbohydrate contents, including starch and total soluble sugar, were significantly increased in response to elevated CO_2. These data should be useful for future gene discovery, molecular studies, and tree improvement strategies for the upcoming increased-CO_2 environments.

Keywords: carbon dioxide; climate change; open-top chamber; RNA sequencing; gene expression analysis; *Populus*

Citation: Kim, T.-L.; Chung, H.; Veerappan, K.; Lee, W.Y.; Park, D.; Lim, H. Physiological and Transcriptome Responses to Elevated CO_2 Concentration in *Populus*. *Forests* **2021**, *12*, 980. https://doi.org/10.3390/f12080980

Academic Editors: Mariangela Fotelli and Cate Macinnis-Ng

Received: 30 March 2021
Accepted: 21 July 2021
Published: 23 July 2021

Publisher's Note: MDPI stays neutral with regard to jurisdictional claims in published maps and institutional affiliations.

1. Introduction

The CO_2 concentration in the atmosphere is rapidly increasing due to industrialization and deforestation, causing global warming and abnormal weather conditions. The reduction in the CO_2 concentration in the atmosphere through photosynthesis highlights the importance of forests (IPCC, 2013). According to the IPCC Fourth Assessment Report (IPCC 2007), the atmospheric CO_2 concentration will increase from 389 ppm in 2005 to 550–700 ppm by 2050 and to 650–1200 ppm by 2100, which will cause global warming of 1.1–6.4 °C by the end of this century [1]. The concentration of CO_2 in the atmosphere directly determines the ratio of plant photosynthesis, and indirectly affects plant productivity and fitness, and thus could serve as a form of selective pressure, though there is little evidence to support this contention. Carbon dioxide is a major component in photosynthesis that converts solar energy into energy stored in carbohydrates, thereby controlling plant yield. An elevated CO_2 concentration raises the photosynthetic rate and plant's net production or biomass [2], but a rising temperature increases the ratio of photorespiratory loss of carbon to photosynthetic gain, thereby having an opposite effect [3]. The increase in

5

the concentration of atmospheric CO_2 is likely to have a significant effect on the photosynthesis, metabolism, and development of plants [4,5]. In plants, increased primary carbon sources accelerate their metabolism and growth, especially when other growth sources are abundant [6,7].

Populus is a significant model for studying the effects of abiotic stresses on trees. In addition, *Populus* is now recognized as a model tree [8], enabling genomic resources to be deployed to answer questions of ecological and evolutionary significance on plant response and adaptation to climate change. The *Populus* species is an excellent model for examining drought stress responses, which affect not only survival but also biomass accumulation [9,10]. Additionally, *Populus* is an economically important wood tree, and in recent years, there has been increasing interest in studying its genotype, transcriptome and drought response mechanisms [11,12]. *Populus* is the internationally accepted model system for physiological and molecular studies in woody plants, in part due to the availability of the complete genome sequence of *Populus trichocarpa* [13]. The availability of the poplar genome plays an important role in understanding the molecular processes of growth, metabolism, and stress responses to environmental changes. The *Populus alba* × *Populus glandulosa* "Clivus" clone is a hybrid species made by NIFoS (National Institute of Forest Science) in Korea. It is a good resource for genetic engineering as it is sterile [14]. The *Populus euramericana* "I-476" clone is an interspecific hybrid produced from the cross of *Populus nigra* and *Populus deltoides*. Many *P. euramericana* clones have been commercialized, used in forestry production, and to promote ecosystem stability [15].

Next-generation sequencing technologies have been used to elucidate the molecular bases of poplar physiological and developmental mechanisms [16,17], and the responses of the *Populus* transcriptome to both biotic and abiotic stresses [18–20]. In particular, RNA-Seq analysis have been an important breakthrough for sensitive, quantitative, annotation-independent, and high-throughput analyses [21].

We aimed to study physiological and transcriptional changes in poplar trees grown under ambient (400 ppm) and elevated (560 and 720 ppm) atmospheric CO_2, and the physiological parameters are correlated with changes in transcript abundance.

2. Materials and Methods

2.1. Plant Materials and Growth Conditions

Poplar seedlings (*Populus alba* × *Populus glandulosa* hybrid "Clivus" clone and *Populus euramericana* "I-476" clone) were grown in pots containing appropriate soil moisture in the open-top chambers (OTCs) system (Figure 1). These *Populus* clones were used as solitary maternal plants. Rooted cuttings from these maternal *Populus* clones were cultivated in pots in a greenhouse for 4 weeks for acclimation and then transferred to OTCs. Ten plants were grown and 8 plants were used in the experiments. The experiment was conducted for 16 weeks in the National Institute of Forest Science in Suwon, Korea (37°15′04″ N, 136°57′59″ E), under natural environmental conditions [22]. Three treatment levels of CO_2 concentration were applied to the OTCs: ambient ($\times 1.0$, ~400 ppm), $\times 1.4$ (~560 ppm), and $\times 1.8$ (~720 ppm). Although the air temperature inside was 1.2–2.0 °C higher than that of the outside, the temperature differences among the OTCs were less than 0.2 °C.

2.2. Measurement of the Chlorophyll Content

The chlorophyll content was determined according to the method of Sibley et al. (1996) [23]. From all treatments, 0.1 g fresh samples were taken in triplicate, homogenized thoroughly with dimethyl formamide (DMF) and centrifuged at $14,000 \times g$ for 10 min at 4 °C. The supernatant was used as the chlorophyll source. The chlorophyll levels were determined by reading the supernatant absorbances at 647 nm and 664 nm with a Biospectrometer (Eppendorf, Hamburg, Germany). The chlorophyll contents and their means were calculated as follows for each plant and treatment:

- Chlorophyll a = $12.7 A_{664} - 2.79 A_{647}$.
- Chlorophyll b = $20.7 A_{647} - 4.62 A_{664}$.

- Carotenoids = $(1000A_{470} - 1.82\text{Chl a} - 85.02\text{Chl b})/198$.

  (A, absorbance; pigment concentration in mg/g fresh weight (FW)).

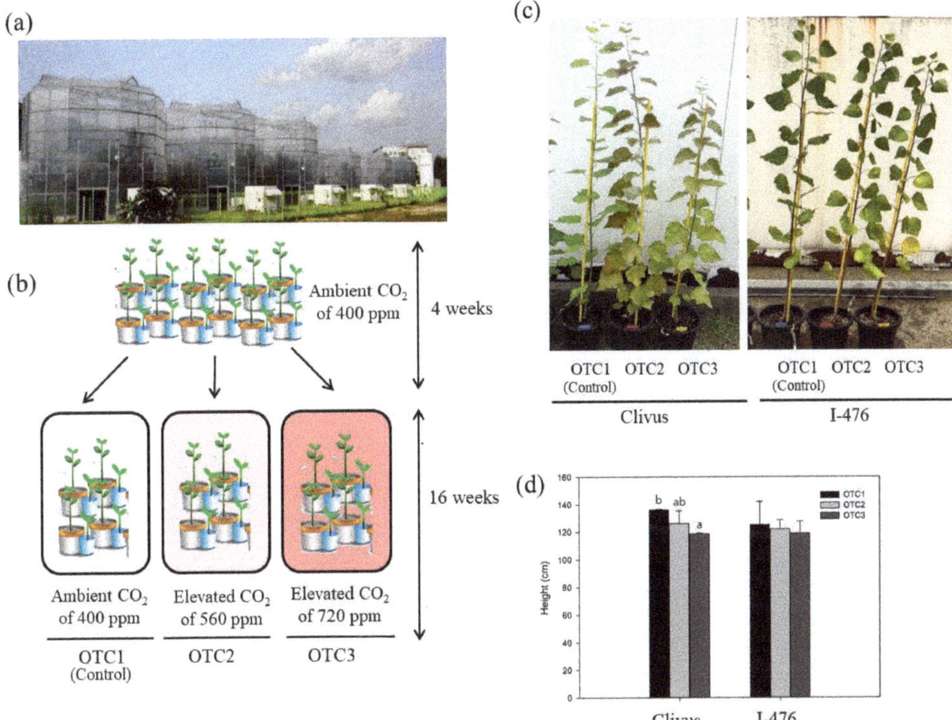

Figure 1. A schematic representation of the experimental design and the phenotypes of poplar clones. (**a**) A photograph of the whole open-top chambers (OTCs) facility. (**b**) The conceptual model. (**c**,**d**) The effects of elevated CO_2 on the shoot growth of the poplar Clivus and I-476 clones. The values are the means ± SDs ($n = 10$) Different lowercase letters indicate significant differences (Tukey's HSD, $p < 0.05$ for (**d**)).

2.3. Extraction and Measurement of Starch and Soluble Sugar

To extract metabolites, approximately 0.1 g poplar leaves from the plants grown for 16 weeks were ground in liquid nitrogen to a fine powder, and then the pulverized tissues were extracted twice with 80% (*v/v*) ethanol at 80 °C. To analyze starch content, the resulting sediments from aqueous ethanol extractions were re-suspended in distilled H_2O and enzymatically digested to glucose according to the method described by Walters et al. (2004) [24]. The sugar concentrations were determined enzymatically by a method described by Stitt et al. (1989) [25] using a Biospectrometer (Eppendorf, Hamburg, Germany).

Total soluble sugars were extracted from leaf tissues by 80% ethanol by a modified method of Irigoyen et al. (1992) [26] as follows. After fresh weight determination, the leaves were homogenized by grinding in liquid nitrogen with mortar and pestle. Then, 2 mL of 80% (*v/v*) ethanol was added and the sample was vortexed for 1 h. After centrifugation at $14,000\times g$ for 10 min, the supernatant was collected. The supernatants were added with chloroform and completely mixed. After centrifugation at $14,000\times g$ for 10 min, 50 µL of supernatant was reacted with 4.95 mL of freshly prepared anthrone reagent (500 mg anthrone + 50 mL 72% H_2SO_4) at 100 °C for 15 min. After cooling on ice, the total soluble sugar content was determined at 620 nm by a Biospectrometer (Eppendorf, Hamburg, Germany) using glucose as the standard.

2.4. Measurement of Malondialdehyde (MDA), Proline and H_2O_2

From each treatment, approximately 0.1 g of poplar leaves were harvested in triplicate and extracted with a buffer consisting of 20% TCA (*w/v*) and 0.5% thiobarbituric acid (TBA) (*w/v*), followed by warming at 95 °C for 30 min. The reaction was terminated by placing the mixture on ice for 30 min and then centrifuging it at 14,000× *g* for 10 min. The absorbance of the supernatant was read at 532 nm using a Biospectrometer (Eppendorf, Hamburg, Germany). The MDA content was derived according to the method of Heath and Packer (1968) [27].

Proline was extracted from a sample of 0.5 g fresh leaf samples in 3% (*w/v*) aqueous sulfosalicylic acid and estimated using the ninhydrin reagent according to the method of Bates et al. (1973) [28]. The absorbance of the fraction with toluene separated from the liquid phase was read at a wavelength of 520 nm. Proline concentration was determined using a calibration curve and expressed as mmol proline g^{-1} FW.

Hydrogen peroxide was measured spectrophotometrically after reaction with KI. The reaction mixture consisted of 0.5 mL 0.1% trichloroacetic acid (TCA), leaf extract supernatant, 0.5 mL of 100 mM K-phosphate buffer and 2 mL reagent (1M KI *w/v* in fresh distilled H_2O). The blank probe consisted of 0.1% TCA, K-phosphate buffer, and KI reagent in the absence of leaf extract. The reaction was developed for 1 h in darkness and absorbance was measured at 390 nm. The amount of hydrogen peroxide was calculated using a standard curve prepared with known concentrations of H_2O_2 [29].

2.5. RNA Isolation, Library Preparation and qRT-PCR Analysis

Samples obtained from the leaves of eight plants in each CO_2 treatment were frozen in liquid nitrogen, grinded, and stored at −80 °C. RNAs from 8 plants for each treatment were extracted, and equal amounts of RNA were pooled to obtain single RNA samples; RNA pools were used for RNA sequencing. cDNAs for qRT-PCR were synthesized from all 48 samples. Total RNA was isolated using a RNeasy plant mini kit (Qiagen, Hilden, Germany). Approximately 2 µg of RNA from each tissue was used to construct cDNA libraries for sequencing according to the Illumina TruSeq RNA Sample Preparation protocol. In short, the workflow included isolation of poly-adenylated RNA molecules using poly-T oligo-attached magnetic beads, enzymatic RNA fragmentation, cDNA synthesis, ligation of bar-coded adapters, and PCR amplification. The libraries were sequenced using an Illumina HiSeq 4000 platform with 101 paired-end sequences at the Macrogen (Korea).

For real-time quantitative RT-PCR analysis, first-strand cDNA was synthesized from 1 µg of DNase-treated total RNA using RNA to cDNA EcoDry^{TM} Premix (TaKaRa, Shiga, Japan). All reactions were performed using the IQ^{tm} SYBR Green Supermix (BIO-RAD, Hercules, CA, USA) and carried out in a CFX96 Touch Real-Time PCR Detection System (BIO-RAD, Hercules, CA, USA) according to the manufacturers' instructions. The gene-specific primers used for the quantitative RT-PCR were found in Supplementary Table S1. The reaction cycle was: 1 cycle of 95 °C for 30 s, followed by 40 cycles of 95 °C for 5 s, and 60 °C for 34 s. Relative quantification was performed to calculate expression levels of target genes in different treatments using the $2^{-\Delta\Delta Ct}$ methods [30]. In the reaction plate, each sample was measured in triplicate. The expression level of *ACTIN1* was used for the normalization of quantitative real-time PCR results [31].

2.6. RNA Sequencing Analysis

The sequenced raw reads from *Populus* young leaf tissues were processed with the following RNA sequencing pipeline. The quality of the RNA-Seq reads from all of the six tissues was checked using FastQC. The Trimmomatic software (v0.0.14) was used to process raw reads to remove adaptor sequences and low quality reads [32]. The obtained high quality reads were mapped into *Populus alba* reference genome (GCA_005239225.1_ ASM523922v1_genomic.fna) from NCBI (https://www.ncbi.nlm.nih.gov/genome/13203, accessed on 15 January 2020) using Hisat2 [33]. Further, SAMtools was implemented to convert SAM files from BAM files [34]. Then, we used Feature Counts [35] to estimate the uniquely mapped gene counts by using *Populus alba* genome annotation (GCA_0052392 25.1_ASM523922v1_ genomic.gff).

2.7. Differentially Expressed Gene (DEG) Analysis

The generated gene count files from each *Populus* tissue (Clivus_OTC1 (C_01), Clivus_OTC2 (C_02), Clivus_OTC3 (C_03), I-476_OTC1 (I_01), I-476_OTC2 (I_02), and I-476_OTC3 (I_03)) were used for differential expression genes analysis using DEseq2 version 1.20 package in R analysis environment [36]. Further, Fisher's exact test and likelihood ratio test methods were implemented to perform differential expression analysis of digital gene expression data following a binomial distribution. This experiment lacks biological replicates and log2 fold change ≥ 2 was used as a threshold to identify the DEGs. Further, only DEGs with a gene symbol were considered for downstream analysis. Common DEGs in both clones from all comparison (400 vs. 560, 400 vs. 720, and 560 vs. 720) were extracted using Venny 2.1.0 for further understanding.

2.8. Annotation of DEGs

The Omicsbox (Blast2GO) program (Biobam, Purchased version, Valencia, Spain) was used to perform functional annotation analysis. In detail, we performed a BLASTx program-based homology search for the *Populus alba* gene sequences against the Arabidopsis protein database (NCBI Arabidopsis protein sequences https://ftp.ncbi.nlm.nih.gov/genomes/all/GCF/000/001/735/GCF_000001735.4_TAIR10.1/GCF_000001735.4_TAIR10.1_protein.faa.gz, accessed on 3 February 2020) using a cutoff E-value $< 10^{-5}$ and a maximum number of allowed hits were fixed at 10 per query. The alignment results with the smallest E-value were considered to select best top hits. Further, GO (Gene Ontology) terms were identified for all genes which are associated from BLASTx (NCBI, https://blast.ncbi.nlm.nih.gov/, accessed on 11 February 2021) search results with Omicsbox (Biobam, Spain, https://www.biobam.com/omicsbox/, accessed on 11 February 2021) and WEGO (Web Gene Ontology Annotation Plot) (http://wego.genomics.cn, accessed on 4 October 2020) used for plotting [37]. The corresponding Arabidopsis gene symbol, along with DEG specific fold values, were used to perform gene set enrichment analysis by WebGestalt (http://www.webgestalt.org/, accessed on 2 May 2020) [38]. Cluster analysis was performed using ClustVis (https://biit.cs.ut.ee/clustvis/, accessed on 11 February 2021) [39].

2.9. Statistical Analysis

Analyses were carried out using a one-way ANOVA with multiple comparisons using Dunnett's T3 or Tukey's HSD. p-values < 0.05 were considered significant. Values are presented as means with SD.

3. Results

3.1. Growth and Physiological Changes in Response to Elevated CO_2

We observed the phenotypes and measured the physiological changes of the two poplars' leaves at the seedling stage. The height of the Clivus grown with an elevated CO_2 level of 720 ppm was significantly reduced as a response to elevated CO_2 (Figure 1). However, except this instance, there were no significant physiological differences observed between ambient or elevated CO_2 poplar plants. The leaves' chlorophyll content was correlated with photosynthetic activity, and the chlorophyll levels were affected by elevated CO_2. The chlorophyll A content of I-476 poplar was significantly reduced at elevated CO_2 levels (560 and 720 ppm). Furthermore, in I-476, reduced chlorophyll: carotenoid content was also observed at elevated CO_2 (720 ppm) compared to the controls (Figure 2). In addition, the total chlorophyll content of I-476 poplar was also significantly reduced at elevated CO_2 levels. In both Clivus and I-476, there was a change in carotenoids at an elevated CO_2 level of 560 ppm. Except for the above-mentioned changes, we found no significant changes in the levels of photosynthetic pigments in poplar grown in any CO_2 conditions.

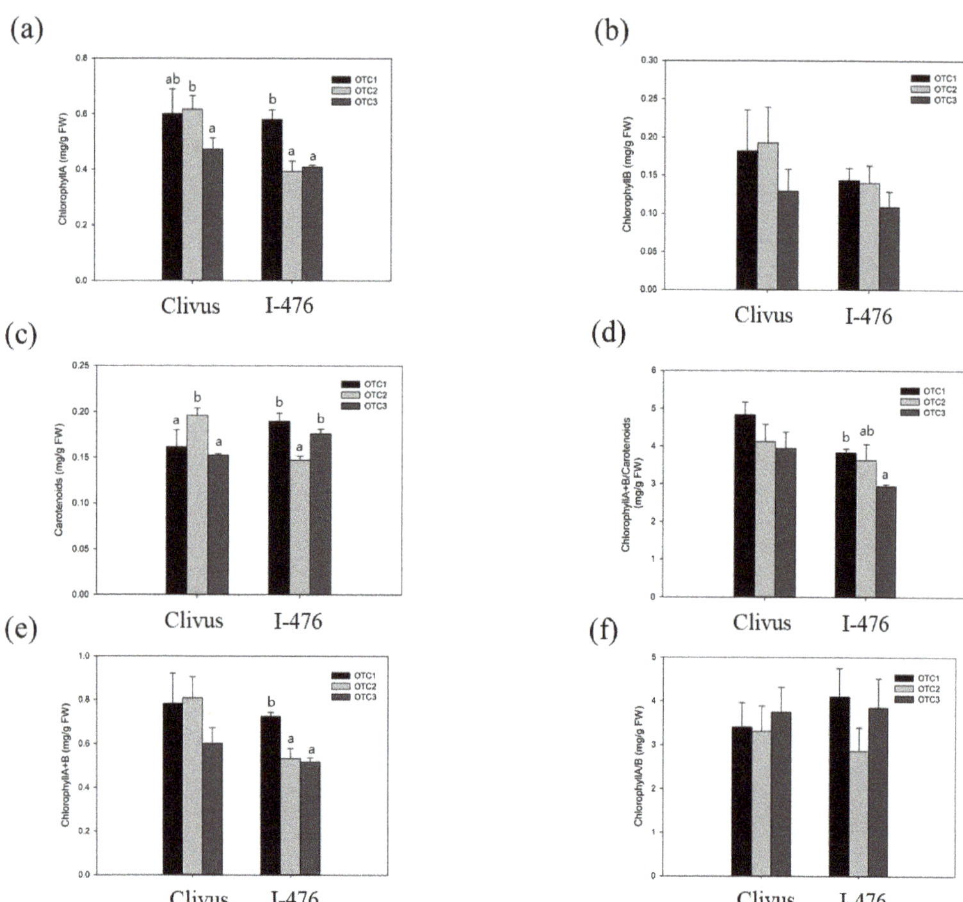

Figure 2. Effects of CO_2 treatment on photosynthetic pigments in poplar clones. (**a**) Chlorophyll A. (**b**) Chlorophyll B. (**c**) Carotenoids. (**d**) Total chlorophyll/carotenoids. (**e**) Total chlorophyll. (**f**) Chlorophyll A/B. Values are means ± SDs of three independent measurements. Different lowercase letters indicate significant differences (Tukey's HSD, $p < 0.05$ for (**a,c,e**); and Welch's ANOVA with Dunnett's T3, $p < 0.05$ for (**d**)).

To investigate the effects of elevated CO_2 on the carbon metabolism and stress-responsive elements, the MDA, H_2O_2, proline, and soluble sugar contents were examined. The MDA content was measured to determine the degree of tissue damage caused by stress [40]. The change in H_2O_2 content is a good indicator of the state of the ability to remove free radicals in oxidation stress [40]. Elevated CO_2 increased the concentrations of MDA, proline, total soluble sugar, and starch in leaves of the Clivus, but did not alter the concentration of H_2O_2. In contrast, elevated CO_2 significantly increased the concentration of H_2O_2, proline, total soluble sugar, and starch in leaves of the I-476. Only the level of MDA at 520 ppm decreased in I-476. In particular, elevated CO_2 tended to increase the concentration of starch in both the Clivus and I-476 (Figure 3).

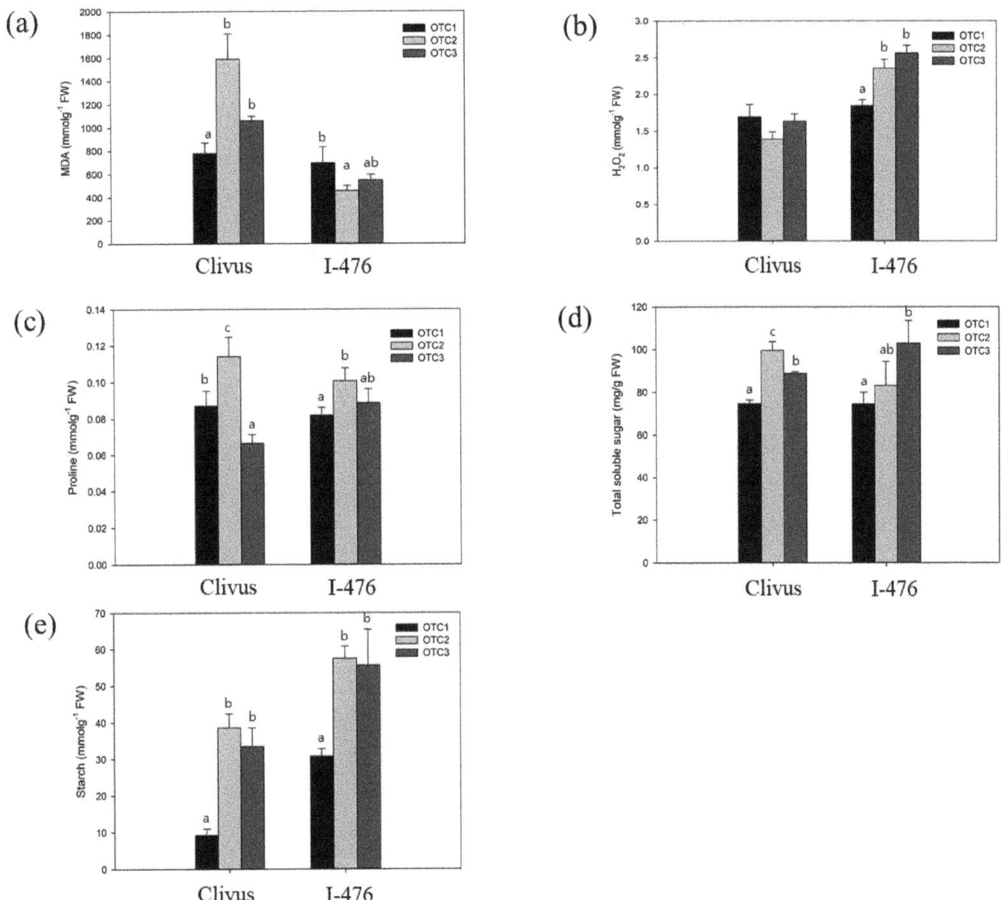

Figure 3. Effects of CO_2 treatment on the metabolites and carbohydrate contents. (**a**) MDA. (**b**) H_2O_2. (**c**) Proline. (**d**) Total soluble sugar. (**e**) Starch. Different lowercase letters indicate significant differences (Tukey's HSD, $p < 0.05$ for I-476 in (**a–e**); and the same for Clivus, except Welch's ANOVA with Dunnett's T3, $p < 0.05$ for Clivus in (**a**)).

3.2. RNA Sequencing Analysis

RNA-Seq analysis was performed in order to study the differential gene expression patterns exerted in response to elevated CO_2 treatments in two poplar species (Clivus and I-476). After the adapter trimming and quality check, the clean reads were mapped to the reference genome *Populus alba*. The mapping percentage varied significantly between the species; Clivus (*P. alba* × *P. glandulosa* hybrid) showed higher percentage matching than the I-476 (*P. euramericana*), as expected. The detailed report on raw read number, data in GB, clean read number and clean read rate for each comparison are detailed in Table 1. The quality score, Q30 clean base rate, was approximately 94% in all comparisons (Table 1).

Table 1. A summary of the RNA sequencing results.

Features	C-O1	C-O2	C-O3	I-O1	I-O2	I-O3
Raw Reads Number	106,478,531	105,000,000	104,645,041	100,369,331	106,750,088	105,451,387
Data in GB	21.5	21.14	21.13	20.27	21.56	21.3
Clean Reads Number	103,924,392	102,000,000	101,893,091	97,491,907	103,843,061	102,166,668
Clean Reads Rate	97.60%	97.58%	97.37%	97.13%	97.27%	96.88%
Low-quality Reads	2.39%	2.41%	2.62%	2.86%	2.72%	3.11%
Total bases	10,750,000,000	10,600,000,000	10,570,000,000	10,137,302,431	10,781,758,888	10,650,590,087
Clean Q30 Bases	10,180,000,000	10,000,000,000	9,903,000,000	9,545,026,030	10,202,693,412	10,386,382,963
Clean Q30 Bases Rate	94.65%	94.69%	93.69%	94.15%	94.62%	93.91%
GC %	44	44	44	44	45	45
Unique Mapped reads	85,670,611	82,924,836	83,342,372	54,226,551	56,729,089	55,987,647
Unique Mapped ratio %	82.44%	81.17%	81.79%	55.62%	54.63%	54.80%
Multi-mapped reads	3,199,788	3,259,580	3,332,247	2,258,383	2,721,419	2,661,064
Multi-mapped reads ratio %	3.08%	3.19%	3.27%	2.32%	2.62%	2.60%
Overall Alignment with Reference Genome	91.08%	89.87%	90.70%	70.41%	69.56%	70.04%

3.3. Differentially Expressed Genes (DEGs)

The DEG analysis was used as an exploratory data to identify the genes responsive to elevated CO_2 treatment (Supplementary Figures S1 and S2). Elevated CO_2 treatment displayed a change in expressed genes in both the clones (Figure 4). Clivus clone showed a greater number of DEGs than I-476 in all comparisons (400 vs. 560, 400 vs. 720, and 560 vs. 720). The higher percentage of mapping observed in Clivus clone to *Populus alba* reference genome might be the reason for the distinct number of DEGs observed between clones. DEGs that were consistently expressed genes in both elevated CO2 treatment could be candidate responsive genes for elevated CO_2 treatment in poplar clones. Therefore, we extracted common DEGs from all comparisons. A total of 29 common genes were observed in Clivus, of which two were upregulated and 24 downregulated (Figure 4). Most of these DEGs belong to the cell wall protein family, especially hydroxyproline-rich glycoproteins (HRGP) [41,42], and the fatty acid metabolism pathway (Figure 5a). HRGP are proline-rich cell wall proteins that have a wide range of functions in signal transduction cascades, such as plant development and stress tolerance. As observed in Figure 5a, the last three genes showed increased or decreased expression in comparison 400 vs. 560, while increasing the CO_2 concentration (400 vs. 720) did not show concomitant increase or decrease. Similarly, in I-476, there were seven genes upregulated and eight genes downregulated in both the elevated CO_2 treatment compared to ambient CO_2 concentration, and the last 10 genes showed varied expression pattern when CO_2 increased. Common DEGs obtained in the I-476 clone belonged to the HSP chaperone family and metabolic pathways, especially photosynthesis, nitrogen metabolism, and carbon metabolism (Figure 5b).

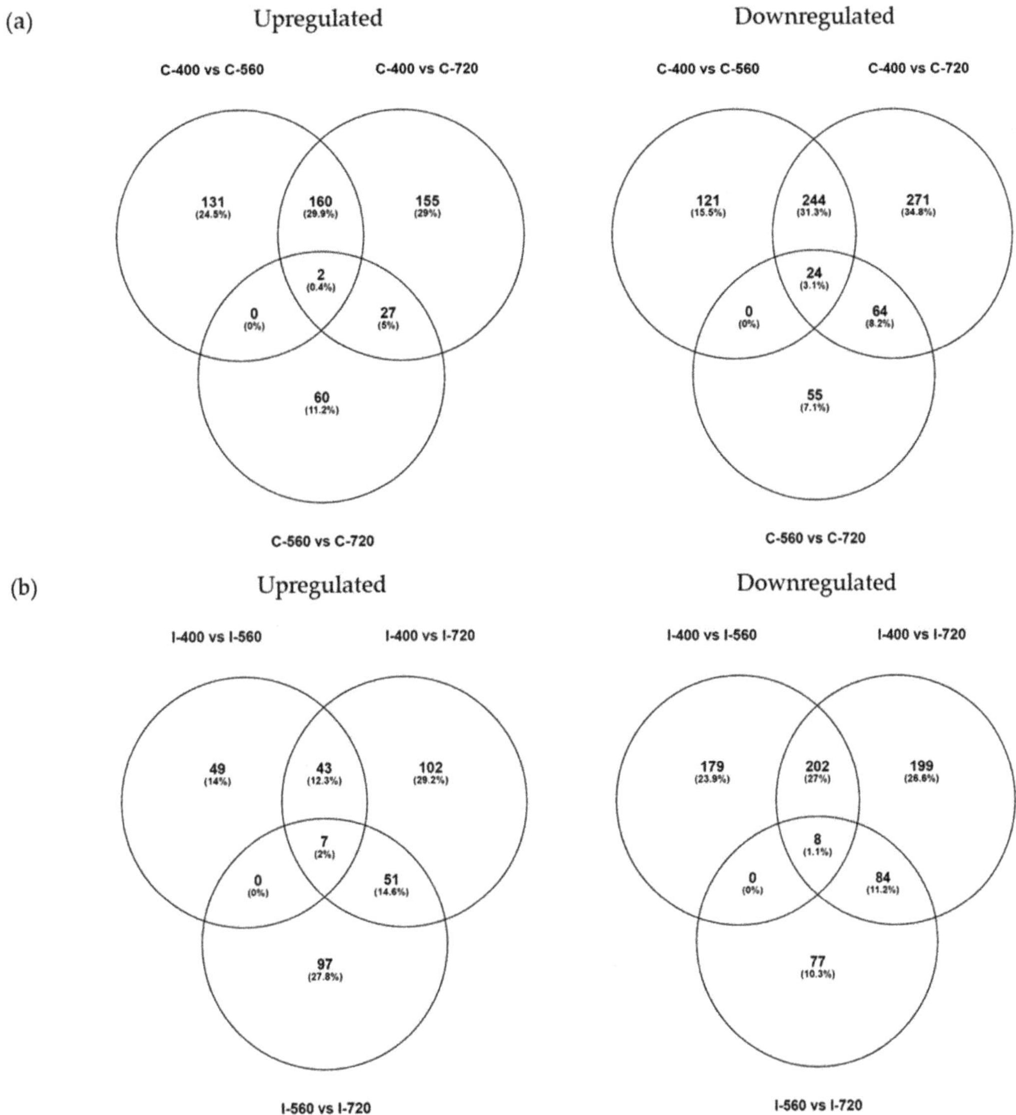

Figure 4. Comparison of differentially expressed genes. (**a**) Clivus (*Populus alba* × *P. glandulosa*, a poplar clone hybrid); and (**b**) I-476 (*Populus euramericana*). Upregulated-log2 fold change (≥ 2); Downregulated-log2 fold change (≤ -2). Each species was treated with three different concentrations of CO_2. C-400, Clivus CO_2 400 ppm; C-560, Clivus CO_2 560 ppm; C-720, Clivus CO_2 720 ppm; I-400, I-476 CO_2 400 ppm; I-560, I-476 CO_2 560 ppm; I-720, I-476 CO_2 720 ppm.

3.4. Functional Classification of Two Populus Species

The GO study displayed a similar pattern of enrichment in both the poplar species. Under the biological process categories, the greater number of DEGs falls under response to stress, responsive to stimulus and external stimulus sub-classes. The CO_2 treatment has triggered the responsive stimulus in the leaves to a greater extent, which was well observed in GO enrichment analysis (Figure 6). Additionally, HGRP and other membrane-associated

DEGs found in elevated CO_2 treated leaves correlated with the plant response in protecting its cell wall.

The KEGG enrichment analysis on DEGs resulted in unanimity pathways in all comparisons, irrespective of species differentiation. Carbon metabolism, starch sucrose metabolism, glutathione metabolism, and fatty acid metabolism are the most common enriched pathways in top 10 lists (Figure 7).

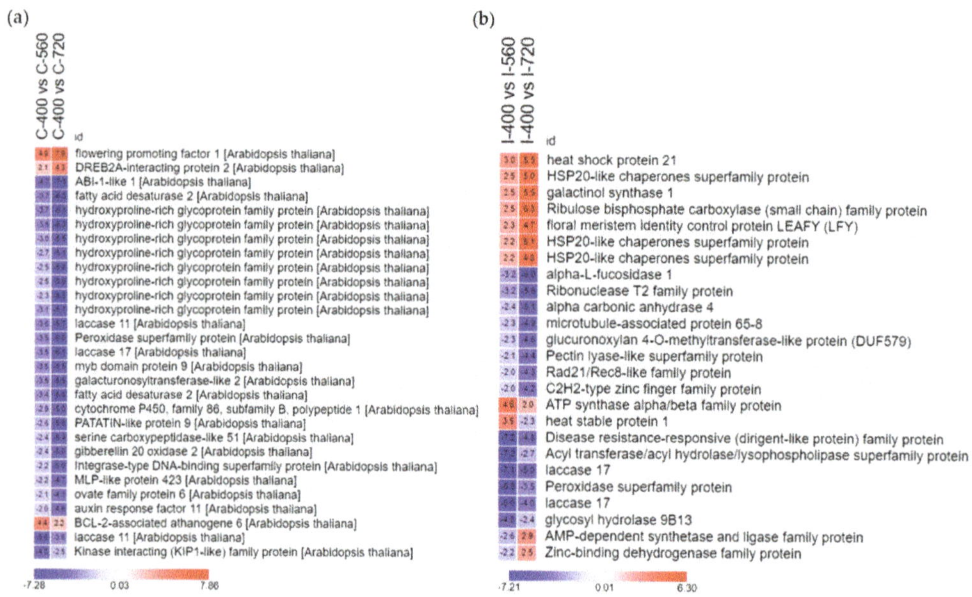

Figure 5. Heat map showing the gene expression pattern of common DEGs in (**a**) Clivus and (**b**) I-476. Corresponding log2 fold change is displayed inside the box. Each species was treated with three different concentrations of CO_2. C-400, Clivus CO_2 400 ppm; C-560, Clivus CO_2 560 ppm; C-720, Clivus CO_2 720 ppm; I-400, I-476 CO_2 400 ppm; I-560, I-476 CO_2 560 ppm; I-720, I-476 CO_2 720 ppm.

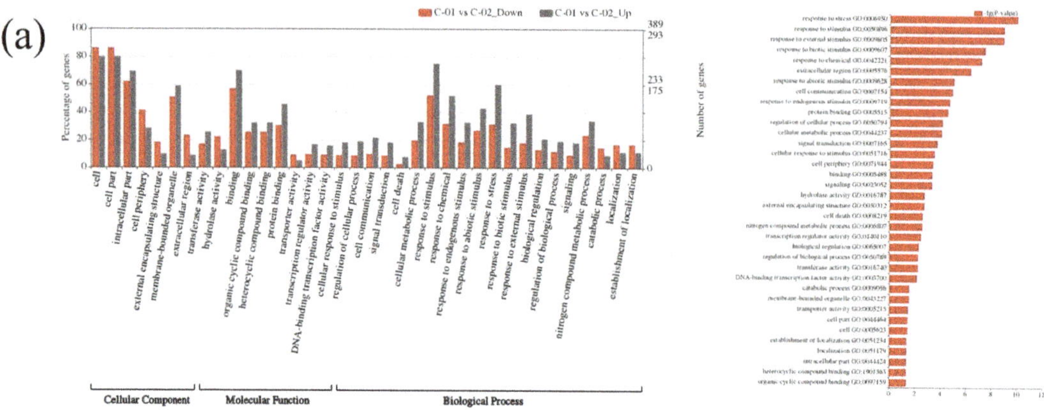

Figure 6. *Cont.*

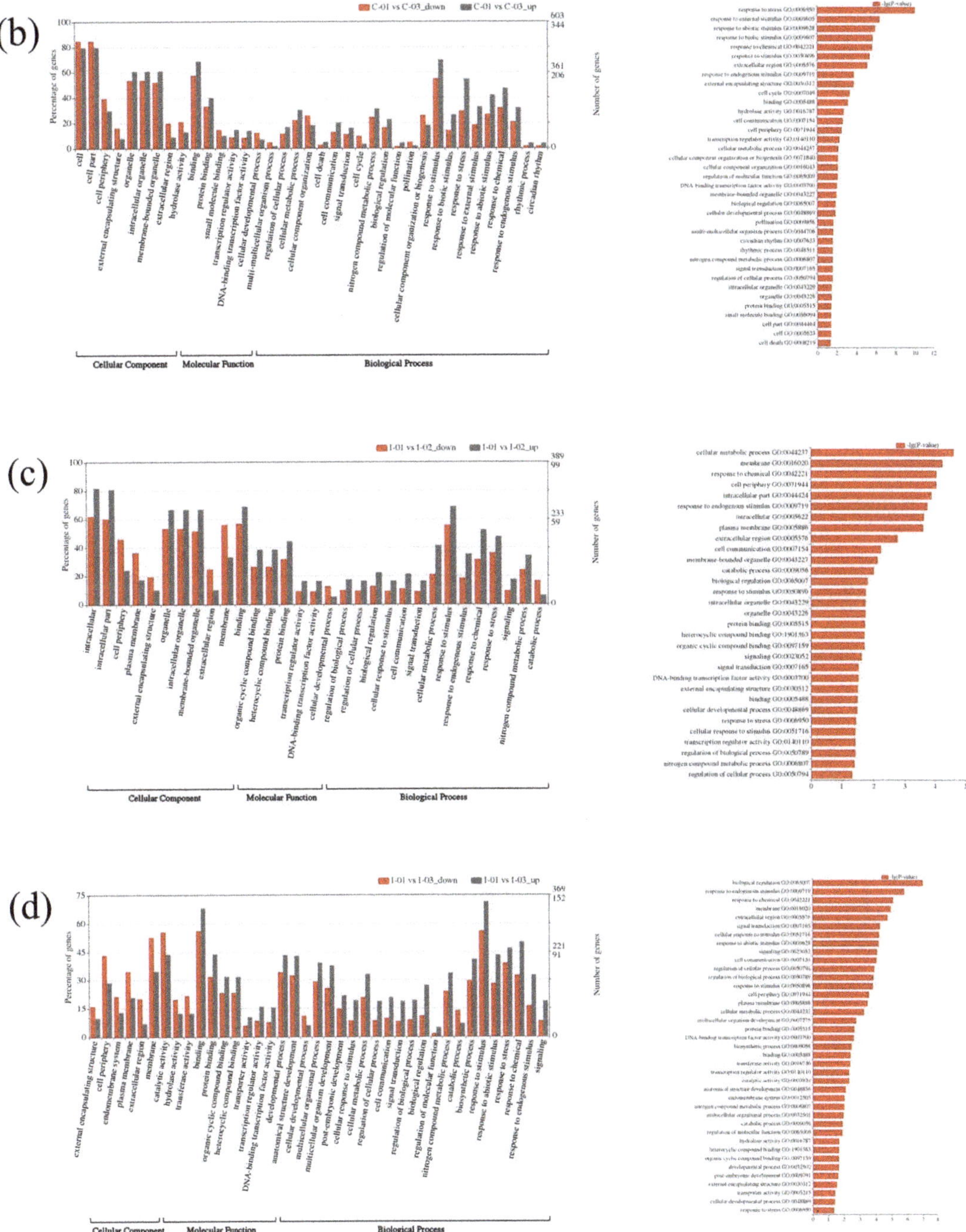

Figure 6. Gene Ontology (GO) enrichment analysis of DEGs. (**a**,**b**) Clivus; (**c**,**d**) I-476. Grey and red represent the upregulated and downregulated DEGs, respectively.

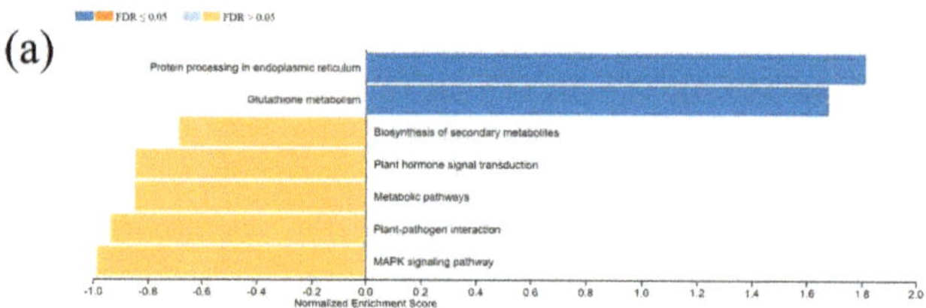

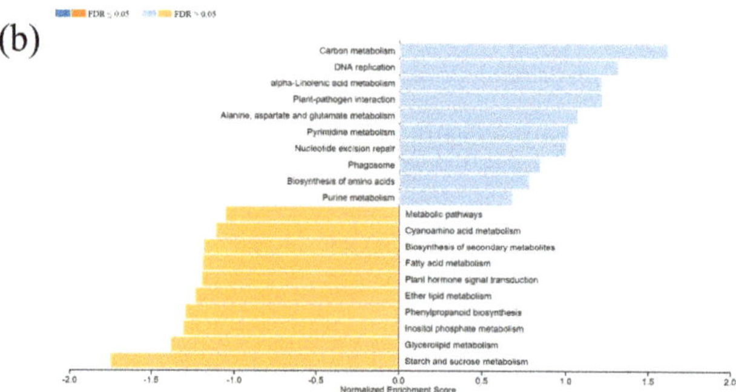

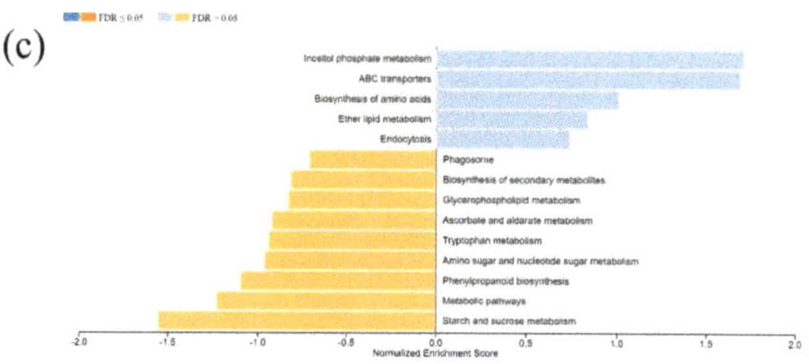

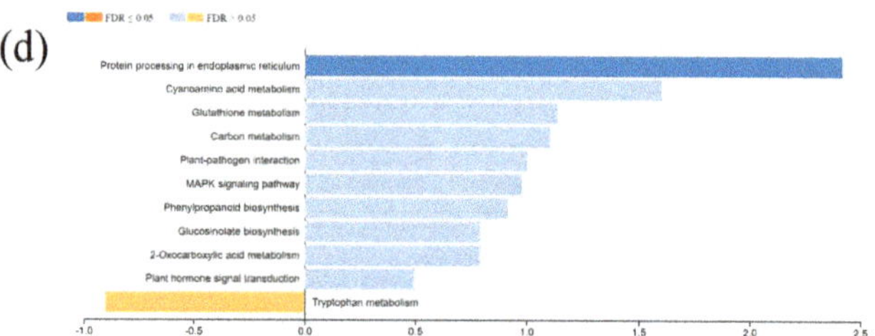

Figure 7. KEGG pathway enrichment analysis. (**a**,**b**) Clivus; (**c**,**d**), I-476.

3.5. Validation of RNA-Seq Results by qRT-PCR

We performed a qRT-PCR for the top 10 most significant DEGs, including D5086_0000265220, D5086_0000080910, D5086_0000065160, D5086_0000242970, D5086_0000065170, D5086_0000289410, D5086_0000064330, D5086_0000042670, D5086_0000250770 and D5086_0000252880 in Clivus (Figure 8a–e,k–o); and D5086_0000176630, D5086_0000043470, D5086_0000080180, D5086_0000215870, D5086_0000038460, D5086_0000212360, D5086_0000328260, D5086_0000281500, D5086_0000305900, and D5086_0000277080 in I-476 (Figure 8f–j,p–t).

To verify the differences in gene expression among OTC1, OTC2, and OTC3, 20 genes were selected for qRT-PCR. In qRT-PCR assay, most of the genes showed similar expression patterns as were observed in RNA-Seq data. We found that common DEGs, D5086_0000080910, D5086_0000065160, D5086_0000242970, and D5086_0000065170, were downregulated, while D5086_0000265220 was upregulated in Clivus species in both OTC2 and OTC3 compared with OTC1. The common DEGs of I-476, except the upregulated D5086_0000038460, D5086_0000176630, D5086_0000043470, D5086_0000080180, and D5086_0000215870, were all downregulated in both OTC2 and OTC3 compared with OTC1. We analyzed the sucrose related genes of in both Clivus and I-476; compared with OTC1, both OTC2 and OTC3 showed upregulation of DEGs D5086_0000064330, D5086_0000042670, D5086_0000250770, and D5086_0000252880, whereas D5086_0000289410 was downregulated in Clivus. In the case of I-476, D5086_0000212360, D5086_0000328260, D5086_0000281500, D5086_0000305900, and D5086_0000277080 were downregulated in both OTC2 and OTC3 compared with OTC1. Among the above results, the decreased expression levels of D5086_0000042670 and D5086_000305900 in OTC2 and OTC3 were opposite to each other when compared with RNA-Seq data.

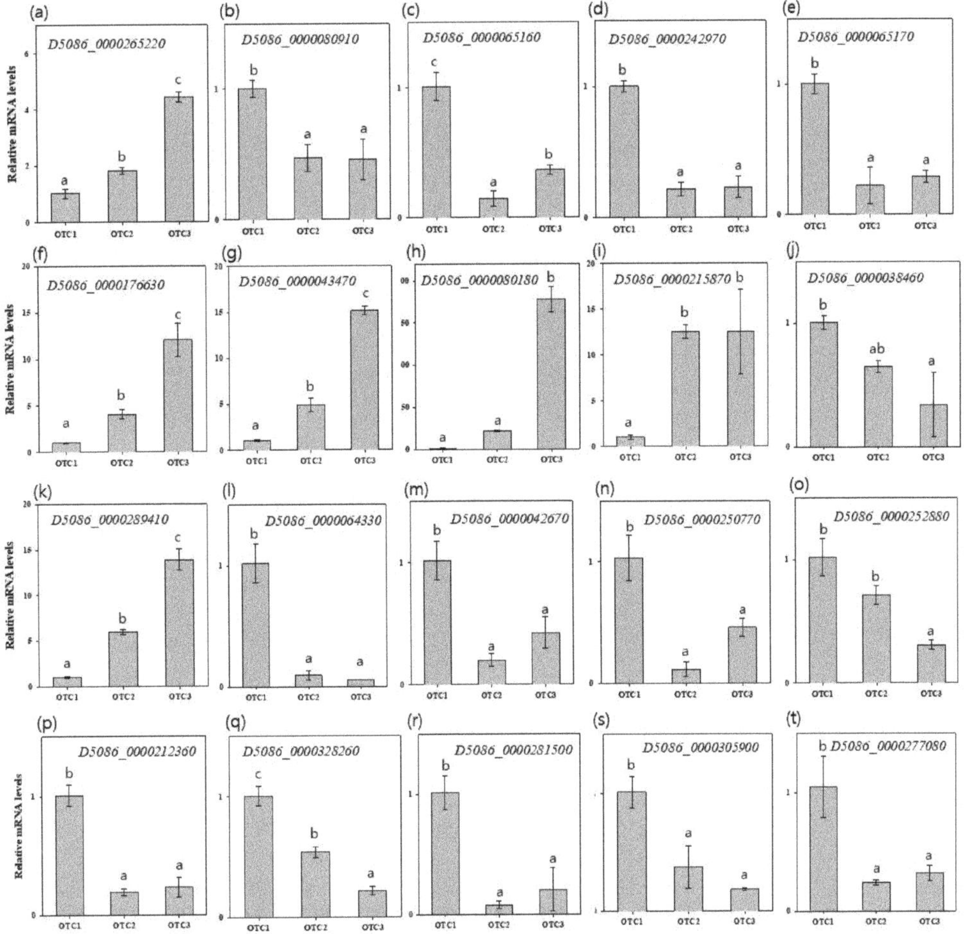

Figure 8. Validation of differential expressions of 20 genes by quantitative real-time PCR (qRT-PCR). qRT-PCR analysis data for the top 10 most significantly differentially expressed genes (DEGs) in Clivus and I-476, respectively. qRT-PCR data were

analyzed using the $2^{-\Delta\Delta Ct}$ method with Actin1 gene as an internal control. Three biological replicates were performed for each sample. Error bars show mean standard error ($n = 3$). (**a–e**) Common Clivus genes; (**f–j**) common I-476 genes; (**k–o**) sucrose genes of Clivus; and (**p–t**) sucrose genes of I-476. Different lowercase letters indicate significant differences. (**a**, DREB2A-interacting protein 2/E3 ubiquitin protein ligase DRIP2-like; **b**, ABI-1-like 1/probable protein ABIL5; **c**, fatty acid desaturase 2/delta(12)-acyl-lipid-desaturase-like; **d**, myb domain protein 9/transcription factor MYB53-like; **e**, fatty acid desaturase 2/delta(12)-acyl-lipid-desaturase-like; **f**, HSP20-like chaperones superfamily protein/23.6 kDa heat shock protein, mitochondrial-like; **g**, galactinol synthase 1/galactinol synthase 2-like; **h**, Ribulose bisphosphate carboxylase (small chain) family/ribulose bisphosphate carboxylase small chain clone 512-like protein; **i**, HSP20-like chaperones superfamily protein/22.0 kDa class IV heat shock protein-like; **j**, glucuronoxylan 4-O-methyltransferase-like protein (DUF579); **k**, Glycogen/starch synthases, ADP-glucose type; **l**, glycosyl hydrolase 9C3; **m**, glycosyl hydrolase 9B13; **n**, glycosyl hydrolase 9B18; **o**, Glycosyl hydrolases family 31 protein; **p**, glycosyl hydrolase 9C3; **q**, Glycosyl hydrolase superfamily protein; **r**, plant glycogenin-like starch initiation protein 1; **s**, Glycosyl hydrolase superfamily protein; and **t**, plant glycogenin-like starch initiation protein 3) (Tukey's HSD, $p < 0.05$ for (**e,f,h,i,k,l,r,t**); and Welch's ANOVA with Dunnett's T, $p < 0.05$ for (**a–d,g,m–q,s**).

4. Discussion

As a rapid increase in CO_2 is expected within decades, studies on the reaction of plants to high concentrations of CO_2 have been steadily progressing. As the initial growing environmental conditions of the seedlings are very important for development and survival, a response study to high concentrations of CO_2 was conducted. The physiological response of poplar to elevated CO_2 varies greatly depending on the species, growth period, additional nutrients, and temperature [7,43–46]. We observed changes in Clivus and I-476 poplar seedlings in OTCs without any additional factor intervention. However, OTCs have an intrinsic constraint in that the temperature (by 1.2–2 °C) is higher than outside due to their limited airflow. Growth deterioration due to spatial constraints of OTCs did not affect growth at the seedling level. Under these conditions, only Clivus clone at an elevated CO_2 level of 720 ppm showed a decrease in growth (Figure 1). In the previous study, impaired plant growth was observed after long-term exposure to a high CO_2 concentration and insufficient nitrogen by altering the primary metabolism [47].

Previous studies have shown that the chlorophyll content may decrease or increase in response to changes in the external environment [48]. In our result, it was confirmed that levels of chlorophyll A and carotenoids showed significant changes (Figure 2). It has been reported that the content of chlorophyll is highly correlated with the concentration of CO_2 in the atmosphere and the photosynthetic ability [49,50]. An elevated CO_2 concentration initially increases the photosynthetic rates of plants, but over time, the rate of photosynthesis is reduced due to the feedback of photosynthetic products and reduction in Rubisco content [51]. Referring to the above, our result showed that treatment with a high concentration of CO_2 for 16 weeks passed the initial stage and reached the mid-term or long-term reaction. In particular, a decrease in chlorophyll A, which is involved in carbon fixation, is suggested by the feedback of accumulated soluble sugar and starch [50,52]. In a recent report, it was shown that in *Populus* grown at high CO_2 concentrations, the content of carotenoids slightly decreased or remained unchanged [53]. This was in accordance with the reduction in the level of carotenoids in our experiment (Figure 2). The nitrogen content of the leaves can decrease and photosynthetic products do not move to the sink organs and accumulate as starch, reducing the rate of photosynthesis [51]. This photosynthetic acclimation is also consistent with the large increase in starch content at elevated CO_2 concentrations in our results (Figure 3). In addition, we observed a decrease in the expression of glycosyl hydrolase genes, which is related to the degradation of starch, and is also consistent with the result of increasing starch content [51,54]. It is well known that an increase in the soluble sugar content in cells inhibits photosynthetic genes [45,55]. In both Clivus and I-476 seedlings we found an increase in the soluble sugar at elevated CO_2 concentrations, and this was associated with a decrease in the level of chlorophyll A in both poplars (Figures 2 and 3). The presence of many DEGs in sucrose and starch metabolic pathways and the related physiological changes observed have confirmed the influential changes in carbon partitioning affecting metabolic changes in poplar under high CO_2 concentration.

Plants grown at elevated CO_2 concentrations have been reported to be more tolerant to drought stress, as stomata can be kept small due to high CO_2 pressure [43]. We confirmed that the overall level of proline increased at elevated CO_2, which is thought to be helpful in improving drought tolerance by affecting the intracellular osmotic pressure regulation of proline (Figure 3). In this study, Clivus showed greater changes in carbon partitioning and stress indicators than I-476. These observations shown in Clivus were operating more pronouncedly at mildly (520 ppm) elevated CO_2 concentrations. As reported previously, H_2O_2 is produced primarily by plant cells during photosynthesis and photorespiration, and affected by environmental stresses [31]. In our results, on the contrary, in I-476, H_2O_2 was increased even though photosynthetic pigment was decreased at an elevated CO_2 concentration. The alterations in the stress-related indicators at elevated CO_2 concentrations might be involved in overlapping stress and elevated CO_2 regulatory networks [56].

Here, we have studied the genetic response to elevated CO_2 on two poplar clones. The transcriptome profiling of poplar clones in elevated CO_2 was utilized as an exploratory data to find the gene response pattern. Due to the lack of statistical testing, transcriptome data from this study has been interpreted with caution. We extracted common DEGs in both clones that are responsive for the elevated CO_2 treatments. The common DEGs from Clivus pooled mainly in HGRP and fatty acid metabolism, whereas for I-476 they were mainly HSPs and from diverse metabolic pathways. This correlates with the experimental data, where in Clivus there is increased MDA production, reflecting the transcriptome analysis showing more DEGs related to fatty acid metabolism. Lesser DEGs related to MDA production were seen in I-476, which explains the lower MDA levels in this clone (Figure 4). The HGRP-related DEGs correlates with the free sugar and proline increase at elevated CO_2 in Clivus [57]. Small HSP maintain ROS homeostasis and cooperate with other stress-related genes in response to diverse biotic and abiotic stresses to protect plant cells and tissues [58]. In I-476, DEG analysis showed an evident HSP chaperone-based defense mechanism involving scavenging of H_2O_2. By treating changes in CO_2 concentration, the stress indicators such as MDA, proline, H_2O_2, and expression of HSP-related genes were changed. In Clivus and I-476, elevated CO_2 concentrations significantly decreased the expression of sucrose synthase (SUS) genes. This reduction in expression level is related to the fact that the regulation of photosynthetic amount was due to photosynthetic acclimation and influenced the synthesis of sucrose, the first product of photosynthesis. The plant glycogenin-like starch initiation protein (PGSIP)-like gene's expression was also decreased in both Clivus and I-476. This seems to suggest that the decrease in starch synthesis genes is related to the reduction in photosynthetic pigments and feedback regulation of accumulated starch in leaves. Although, the transcriptome analysis throws light on increased CO_2 concentration induced gene expression changes in these poplar clones, the absence of biological replicates in transcriptome analysis warrants further studies for increased specificity.

5. Conclusions

To summarize, our results present a transcriptome dataset and physiological analysis, which contributes to the understanding of gene expression profiling in response to elevated CO_2. By comparing the transcriptome data of three different concentrations of CO_2, we found that many DEGs were identified as unique to Clivus and I-476. In addition, this study presented significant physiological alterations with respect to chlorophyll A and carotenoids. Additionally, we reported alterations in starch and soluble sugar. Several synthesis and degradation-related sucrose and starch genes showed altered expression dynamics in our transcriptome and qRT-PCR analysis. The transcriptome data and physiological response of poplars at elevated CO_2 presented here could serve as a resource for further research on poplar trees in different CO_2 environments.

Supplementary Materials: The following are available online at https://www.mdpi.com/article/10.3390/f12080980/s1, Figure S1: MA plot of DEGs, Figure S2: Volcano plot of DEGs, Table S1: Primer of qRT-PCR in Clivus and I-476.

Author Contributions: Conceptualization, H.L. and T.-L.K.; methodology, H.L.; software, H.C.; validation, T.-L.K and D.P.; formal analysis, K.V.; investigation, T.-L.K.; resources, W.Y.L.; data curation, H.C.; writing—Original draft preparation, T.-L.K.; writing—Review and editing, H.L.; visualization, T.-L.K.; and supervision, H.L. All authors have read and agreed to the published version of the manuscript.

Funding: This research received no external funding.

Data Availability Statement: Data is contained within the article or Supplementary Materials.

Conflicts of Interest: The authors declare no conflict of interest.

References

1. Solomon, S.; Qin, D.; Manning, M.; Chen, Z.; Marquis, M.; Averyt, K.; Tignor, M.; Miller, H. IPCC Fourth Assessment Report (AR4). In *Climate Change*; IPCC: Geneva, Switzerland, 2007; pp. 133–171.
2. Leakey, A.D.; Ainsworth, E.A.; Bernacchi, C.J.; Rogers, A.; Long, S.P.; Ort, D.R. Elevated CO_2 effects on plant carbon, nitrogen, and water relations: Six important lessons from FACE. *J. Exp. Bot.* **2009**, *60*, 2859–2876. [CrossRef] [PubMed]
3. Long, S. Modification of the response of photosynthetic productivity to rising temperature by atmospheric CO_2 concentrations: Has its importance been underestimated? *Plant Cell Environ.* **1991**, *14*, 729–739. [CrossRef]
4. Ainsworth, E.A.; Long, S.P. What have we learned from 15 years of free-air CO_2 enrichment (FACE)? A meta-analytic review of the responses of photosynthesis, canopy properties and plant production to rising CO_2. *New Phytol.* **2005**, *165*, 351–372. [CrossRef]
5. Nowak, R.S.; Ellsworth, D.S.; Smith, S.D. Functional responses of plants to elevated atmospheric CO_2—Do photosynthetic and productivity data from FACE experiments support early predictions? *New Phytol.* **2004**, *162*, 253–280. [CrossRef]
6. Körner, C. Plant CO_2 responses: An issue of definition, time and resource supply. *New Phytol.* **2006**, *172*, 393–411. [CrossRef]
7. Lindroth, R.L. Impacts of elevated atmospheric CO_2 and O_3 on forests: Phytochemistry, trophic interactions, and ecosystem dynamics. *J. Chem. Ecol.* **2010**, *36*, 2–21. [CrossRef]
8. Jansson, S.; Douglas, C.J. Populus: A model system for plant biology. *Annu. Rev. Plant Biol.* **2007**, *58*, 435–458. [CrossRef]
9. Marron, N.; Delay, D.; Petit, J.-M.; Dreyer, E.; Kahlem, G.; Delmotte, F.M.; Brignolas, F. Physiological traits of two *Populus* × *euramericana* clones, Luisa Avanzo and Dorskamp, during a water stress and re-watering cycle. *Tree Physiol.* **2002**, *22*, 849–858. [CrossRef]
10. Monclus, R.; Dreyer, E.; Villar, M.; Delmotte, F.M.; Delay, D.; Petit, J.M.; Barbaroux, C.; Le Thiec, D.; Bréchet, C.; Brignolas, F. Impact of drought on productivity and water use efficiency in 29 genotypes of *Populus deltoides* × *Populus nigra*. *New Phytol.* **2006**, *169*, 765–777. [CrossRef]
11. Caruso, A.; Chefdor, F.; Carpin, S.; Depierreux, C.; Delmotte, F.M.; Kahlem, G.; Morabito, D. Physiological characterization and identification of genes differentially expressed in response to drought induced by PEG 6000 in Populus canadensis leaves. *J. Plant Physiol.* **2008**, *165*, 932–941. [CrossRef]
12. Song, Y.; Wang, Z.; Bo, W.; Ren, Y.; Zhang, Z.; Zhang, D. Transcriptional profiling by cDNA-AFLP analysis showed differential transcript abundance in response to water stress in Populus hopeiensis. *BMC Genom.* **2012**, *13*, 1–18. [CrossRef]
13. Tuskan, G.A.; Difazio, S.; Jansson, S.; Bohlmann, J.; Grigoriev, I.; Hellsten, U.; Putnam, N.; Ralph, S.; Rombauts, S.; Salamov, A. The genome of black cottonwood, Populus trichocarpa (Torr. & Gray). *Science* **2006**, *313*, 1596–1604. [PubMed]
14. Park, J.; Kim, Y.; Xi, H.; Kwon, W.; Kwon, M. The complete chloroplast and mitochondrial genomes of Hyunsasi tree, *Populus alba* × *Populus glandulosa* (Salicaceae). *Mitochondrial DNA Part B* **2019**, *4*, 2521–2522. [CrossRef] [PubMed]
15. Cui, H.-Y.; Lee, H.-S.; Oh, C.-Y.; Han, S.-H.; Lee, K.-J.; Lee, H.-J.; Kang, K.-S.; Park, S.-Y. High-frequency regeneration by stem disc culture in selected clones of Populus euramericana. *J. Plant Biotechnol.* **2014**, *41*, 236–241. [CrossRef]
16. Dharmawardhana, P.; Brunner, A.M.; Strauss, S.H. Genome-wide transcriptome analysis of the transition from primary to secondary stem development in Populus trichocarpa. *BMC Genom.* **2010**, *11*, 1–19. [CrossRef] [PubMed]
17. Schrader, J.; Nilsson, J.; Mellerowicz, E.; Berglund, A.; Nilsson, P.; Hertzberg, M.; Sandberg, G. A high-resolution transcript profile across the wood-forming meristem of poplar identifies potential regulators of cambial stem cell identity. *Plant Cell* **2004**, *16*, 2278–2292. [CrossRef] [PubMed]
18. Berta, M.; Giovannelli, A.; Sebastiani, F.; Camussi, A.; Racchi, M. Transcriptome changes in the cambial region of poplar (*Populus alba* L.) in response to water deficit. *Plant Biol.* **2010**, *12*, 341–354. [CrossRef]
19. Chen, S.; Jiang, J.; Li, H.; Liu, G. The salt-responsive transcriptome of *Populus simonii* × *Populus nigra* via DGE. *Gene* **2012**, *504*, 203–212. [CrossRef]
20. Ralph, S.G.; Chun, H.J.E.; Cooper, D.; Kirkpatrick, R.; Kolosova, N.; Gunter, L.; Tuskan, G.A.; Douglas, C.J.; Holt, R.A.; Jones, S.J. Analysis of 4,664 high-quality sequence-finished poplar full-length cDNA clones and their utility for the discovery of genes responding to insect feeding. *BMC Genom.* **2008**, *9*, 1–18. [CrossRef]

21. Metzker, M.L. Sequencing technologies—The next generation. *Nat. Rev. Genet.* **2010**, *11*, 31–46. [CrossRef]
22. Lee, J.-C.; Kim, D.-H.; Kim, G.-N.; Kim, P.-G.; Han, S.-H. Long-term Climate Change Research Facility for Trees: CO_2-Enriched Open Top Chamber System. *Korean J. Agric. For. Meteorol.* **2012**, *14*, 19–27. [CrossRef]
23. Sibley, J.L.; Eakes, D.J.; Gilliam, C.H.; Keever, G.J.; Dozier, W.A.; Himelrick, D.G. Foliar SPAD-502 m values, nitrogen levels, and extractable chlorophyll for red maple selections. *HortScience* **1996**, *31*, 468–470. [CrossRef]
24. Walters, R.G.; Ibrahim, D.G.; Horton, P.; Kruger, N.J. A mutant of Arabidopsis lacking the triose-phosphate/phosphate translocator reveals metabolic regulation of starch breakdown in the light. *Plant Physiol.* **2004**, *135*, 891–906. [CrossRef] [PubMed]
25. Stitt, M.; Lilley, R.M.; Gerhardt, R.; Heldt, H.W. [32] Metabolite levels in specific cells and subcellular compartments of plant leaves. *Methods Enzymol.* **1989**, *174*, 518–552.
26. Irigoyen, J.; Einerich, D.; Sánchez-Díaz, M. Water stress induced changes in concentrations of proline and total soluble sugars in nodulated alfalfa (Medicago sativd) plants. *Physiol. Plant.* **1992**, *84*, 55–60. [CrossRef]
27. Heath, R.L.; Packer, L. Photoperoxidation in isolated chloroplasts: I. Kinetics and stoichiometry of fatty acid peroxidation. *Arch. Biochem. Biophys.* **1968**, *125*, 189–198. [CrossRef]
28. Bates, L.S.; Waldren, R.P.; Teare, I. Rapid determination of free proline for water-stress studies. *Plant Soil* **1973**, *39*, 205–207. [CrossRef]
29. Alexieva, V.; Sergiev, I.; Mapelli, S.; Karanov, E. The effect of drought and ultraviolet radiation on growth and stress markers in pea and wheat. *Plant Cell Environ.* **2001**, *24*, 1337–1344. [CrossRef]
30. Livak, K.J.; Schmittgen, T.D. Analysis of relative gene expression data using real-time quantitative PCR and the $2^{-\Delta\Delta CT}$ method. *Methods* **2001**, *25*, 402–408. [CrossRef] [PubMed]
31. Zhang, W.; Chu, Y.; Ding, C.; Zhang, B.; Huang, Q.; Hu, Z.; Huang, R.; Tian, Y.; Su, X. Transcriptome sequencing of transgenic poplar (*Populus × euramericana* 'Guariento') expressing multiple resistance genes. *BMC Genet.* **2014**, *15*, S7. [CrossRef] [PubMed]
32. Bolger, A.M.; Lohse, M.; Usadel, B. Trimmomatic: A flexible trimmer for Illumina sequence data. *Bioinformatics* **2014**, *30*, 2114–2120. [CrossRef] [PubMed]
33. Kim, D.; Paggi, J.M.; Park, C.; Bennett, C.; Salzberg, S.L. Graph-based genome alignment and genotyping with HISAT2 and HISAT-genotype. *Nat. Biotechnol.* **2019**, *37*, 907–915. [CrossRef] [PubMed]
34. Li, H.; Handsaker, B.; Wysoker, A.; Fennell, T.; Ruan, J.; Homer, N.; Marth, G.; Abecasis, G.; Durbin, R. The sequence alignment/map format and SAMtools. *Bioinformatics* **2009**, *25*, 2078–2079. [CrossRef]
35. Liao, Y.; Smyth, G.K.; Shi, W. featureCounts: An efficient general purpose program for assigning sequence reads to genomic features. *Bioinformatics* **2014**, *30*, 923–930. [CrossRef] [PubMed]
36. Love, M.I.; Huber, W.; Anders, S. Moderated estimation of fold change and dispersion for RNA-seq data with DESeq2. *Genome Biol.* **2014**, *15*, 1–21. [CrossRef]
37. Ye, J.; Zhang, Y.; Cui, H.; Liu, J.; Wu, Y.; Cheng, Y.; Xu, H.; Huang, X.; Li, S.; Zhou, A. WEGO 2.0: A web tool for analyzing and plotting GO annotations, 2018 update. *Nucleic Acids Res.* **2018**, *46*, W71–W75. [CrossRef] [PubMed]
38. Liao, Y.; Wang, J.; Jaehnig, E.J.; Shi, Z.; Zhang, B. WebGestalt 2019: Gene set analysis toolkit with revamped UIs and APIs. *Nucleic Acids Res.* **2019**, *47*, W199–W205. [CrossRef]
39. Metsalu, T.; Vilo, J. ClustVis: A web tool for visualizing clustering of multivariate data using Principal Component Analysis and heatmap. *Nucleic Acids Res.* **2015**, *43*, W566–W570. [CrossRef]
40. Kim, J.-S.; Shim, I.-S.; Kim, M.-J. Physiological response of Chinese cabbage to salt stress. *Hortic. Sci. Technol.* **2010**, *28*, 343–352.
41. Kavi Kishor, P.B.; Hima Kumari, P.; Sunita, M.; Sreenivasulu, N. Role of proline in cell wall synthesis and plant development and its implications in plant ontogeny. *Front. Plant Sci.* **2015**, *6*, 544. [CrossRef]
42. Showalter, A.M.; Keppler, B.D.; Liu, X.; Lichtenberg, J.; Welch, L.R. Bioinformatic identification and analysis of hydroxyproline-rich glycoproteins in Populus trichocarpa. *BMC Plant Biol.* **2016**, *16*, 1–34. [CrossRef]
43. Calfapietra, C.; Tulva, I.; Eensalu, E.; Perez, M.; De Angelis, P.; Scarascia-Mugnozza, G.; Kull, O. Canopy profiles of photosynthetic parameters under elevated CO_2 and N fertilization in a poplar plantation. *Environ. Pollut.* **2005**, *137*, 525–535. [CrossRef]
44. Dickson, R.; Coleman, M.D.; Riemenschneider, D.; Isebrands, J.; Hogan, G.; Karnosky, D. Growth of five hybrid poplar genotypes exposed to interacting elevated CO_2 and O_3. *Can. J. For. Res.* **1998**, *28*, 1706–1716. [CrossRef]
45. Ceulemans, R.; Taylor, G.; Bosac, C.; Wilkins, D.; Besford, R. Photosynthetic acclimation to elevated CO_2 in poplar grown in glasshouse cabinets or in open top chambers depends on duration of exposure. *J. Exp. Bot.* **1997**, *48*, 1681–1689. [CrossRef]
46. Marinari, S.; Calfapietra, C.; De Angelis, P.; Mugnozza, G.S.; Grego, S. Impact of elevated CO_2 and nitrogen fertilization on foliar elemental composition in a short rotation poplar plantation. *Environ. Pollut.* **2007**, *147*, 507–515. [CrossRef] [PubMed]
47. Takatani, N.; Ito, T.; Kiba, T.; Mori, M.; Miyamoto, T.; Maeda, S.-I.; Omata, T. Effects of high CO_2 on growth and metabolism of Arabidopsis seedlings during growth with a constantly limited supply of nitrogen. *Plant Cell Physiol.* **2014**, *55*, 281–292. [CrossRef] [PubMed]
48. Zhao, C.; Liu, Q. Growth and photosynthetic responses of two coniferous species to experimental warming and nitrogen fertilization. *Can. J. For. Res.* **2009**, *39*, 1–11. [CrossRef]
49. Terashima, I.; Evans, J.R. Effects of light and nitrogen nutrition on the organization of the photosynthetic apparatus in spinach. *Plant Cell Physiol.* **1988**, *29*, 143–155.
50. Hikosaka, K.; Terashima, I. A model of the acclimation of photosynthesis in the leaves of C3 plants to sun and shade with respect to nitrogen use. *Plant Cell Environ.* **1995**, *18*, 605–618. [CrossRef]

51. Davey, P.; Olcer, H.; Zakhleniuk, O.; Bernacchi, C.; Calfapietra, C.; Long, S.; Raines, C. Can fast-growing plantation trees escape biochemical down-regulation of photosynthesis when grown throughout their complete production cycle in the open air under elevated carbon dioxide? *Plant Cell Environ.* **2006**, *29*, 1235–1244. [CrossRef] [PubMed]

52. Evans, J.R. Partitioning of nitrogen between and within leaves grown under different irradiances. *Funct. Plant Biol.* **1989**, *16*, 533–548. [CrossRef]

53. Lee, S.; Oh, C.-Y.; Han, S.-H.; Kim, K.W.; Kim, P.-G. Photosynthetic Responses of Populus alba× glandulosa to Elevated CO_2 Concentration and Air Temperature. *Korean J. Agric. For. Meteorol.* **2014**, *16*, 22–28. [CrossRef]

54. Tallis, M.; Lin, Y.; Rogers, A.; Zhang, J.; Street, N.; Miglietta, F.; Karnosky, D.; De Angelis, P.; Calfapietra, C.; Taylor, G. The transcriptome of Populus in elevated CO_2 reveals increased anthocyanin biosynthesis during delayed autumnal senescence. *New Phytol.* **2010**, *186*, 415–428. [CrossRef] [PubMed]

55. Nie, G.; Hendrix, D.L.; Webber, A.N.; Kimball, B.A.; Long, S.P. Increased accumulation of carbohydrates and decreased photosynthetic gene transcript levels in wheat grown at an elevated CO_2 concentration in the field. *Plant Physiol.* **1995**, *108*, 975–983. [CrossRef] [PubMed]

56. Palit, P.; Ghosh, R.; Tolani, P.; Tarafdar, A.; Chitikineni, A.; Bajaj, P.; Sharma, M.; Kudapa, H.; Varshney, R.K. Molecular and Physiological Alterations in Chickpea under Elevated CO_2 Concentrations. *Plant Cell Physiol.* **2020**, *61*, 1449–1463. [CrossRef]

57. Abdel-Nasser, L.; Abdel-Aal, A. Effect of elevated CO_2 and drought on proline metabolism and growth of safflower (Carthamus mareoticus L.) seedlings without improving water status. *Pak. J. Biol. Sci.* **2002**, *5*, 523–528.

58. Yer, E.N.; Baloglu, M.C.; Ziplar, U.T.; Ayan, S.; Unver, T. Drought-responsive Hsp70 gene analysis in populus at genome-wide level. *Plant Mol. Biol. Rep.* **2016**, *34*, 483–500. [CrossRef]

forests

Article

Field Performances of Mediterranean Oaks in Replicate Common Gardens for Future Reforestation under Climate Change in Central and Southern Europe: First Results from a Four-Year Study

Filippos Bantis [1,*], Julia Graap [1], Elena Früchtenicht [1], Filippo Bussotti [2], Kalliopi Radoglou [3] and Wolfgang Brüggemann [1,4]

[1] Department of Ecology, Evolution and Diversity, Goethe University Frankfurt, Max-von-Laue-Str.13, D-60438 Frankfurt, Germany; julia.graap@hotmail.de (J.G.); e.fruechtenicht@gmx.de (E.F.); w.brueggemann@bio.uni-frankfurt.de (W.B.)

[2] Department of Agri-Food Production and Environmental Science, University of Florence, Piazzale delle Cascine 28, I-50144 Florence, Italy; filippo.bussotti@unifi.it

[3] Department of Forestry and Management of the Environment and Natural Resources, Democritus University of Thrace, Pantazidou 193, GR-68200 Nea Orestiada, Greece; kradoglo@fmenr.duth.gr

[4] Senckenberg Biodiversity and Climate Research Center, Senckenberganlage 25, D-60325 Frankfurt, Germany

* Correspondence: fbanths@gmail.com

Citation: Bantis, F.; Graap, J.; Früchtenicht, E.; Bussotti, F.; Radoglou, K.; Brüggemann, W. Field Performances of Mediterranean Oaks in Replicate Common Gardens for Future Reforestation under Climate Change in Central and Southern Europe: First Results from a Four-Year Study. *Forests* **2021**, *12*, 678. https://doi.org/10.3390/f12060678

Academic Editor: Carlos Gonzalez-Benecke

Received: 2 March 2021
Accepted: 24 May 2021
Published: 26 May 2021

Abstract: Climate change imposes severe stress on European forests, with forest degradation already visible in several parts of Europe. Thus adaptation of forestry applications in Mediterranean areas and central Europe is necessary. Proactive forestry management may include the planting of Mediterranean oak species in oak-bearing Central European regions. Five replicate common gardens of Greek and Italian provenances of *Quercus ilex*, *Q. pubescens* and *Q. frainetto* seedlings (210 each per plantation) were established in Central Italy, NE Greece (two) and Southern Germany (two, including *Q. robur*) to assess their performance under different climate conditions. Climate and soil data of the plantation sites are given and seedling establishment was monitored for survival and morphological parameters. After 3 years (2019) survival rates were satisfactory in the German and Italian sites, whereas the Greek sites exerted extremely harsh conditions for the seedlings, including extreme frost and drought events. In Germany, seedlings suffered extreme heat and drought periods in 2018 and 2019 but responded well. Provenances were ranked for each country for their performance after plantation. In Greece and Italy, *Q. pubescens* was the best performing species. In Germany, *Q. pubescens* and *Q. robur* performed best. We suggest that Greek or Italian provenances of *Q. pubescens* may be effectively used for future forestation purposes in Central Europe. For the establishment of *Quercus* plantations in Northern Greece, irrigation appears to be a crucial factor in seedling establishment.

Keywords: *Quercus*; morphology evaluation; survival rate; extreme frost; heat and drought

1. Introduction

Climate change scenarios bring forward particularly challenging effects on European forests leading to the necessary development of new strategies for the maintenance of functional forests. Extreme temperatures and severe drought events are predicted to increase in frequency, like it has previously been recorded in 2018 and 2019 across Europe [1,2]. Environmental changes can be so abrupt that natural replacement of forest tree species with better adapted species may not occur at sufficient speed to prevent severe forest degradation. Southern regions may lose large areas of their forest cover altogether, except for their higher mountain ranges. The total net financial loss of the value of European forest land from these shifts is estimated to be up to hundreds of billions Euro in the period to 2100 [3], while ecological implications are also substantial.

23

Proactive forestry management may include the planting of Mediterranean oak species in oak (*Quercus robur* L., *Q. petraea* Matt.) bearing Central European regions [4,5]. This approach aims at the establishment of (at least) seed donor and (at best) timber plantations of future climate-tolerant species within existing oak stands to ensure the continuous presence of closely related oak species, thus maintaining oak-dominated ecosystem functions including oak-associated biodiversity. Such a concept is currently realized e.g., in the South Hesse Oak Project (SHOP [5]). However, since Central Europe and low/medium elevation forests in the Mediterranean will face–in principle–the same problems in the forthcoming century, this approach may be extended to forest management strategies in mountain regions in Mediterranean countries as well.

One way of testing the response of species or populations under certain environmental conditions is by establishing common gardens in several locations of ecological interest [6]. Common gardens are considered essential in determining the potential of selected species and/or provenances [7] to be implemented for 'assisted migration' due to climate change [8–10].

For reforestation purposes, different strategies can be applied. One option is to start the process by active seed dispersal, usually leading to healthy plants with well-developed root systems, especially in tap rooting species like most oaks [11]. However, buried seeds are often prone to herbivory in the field (e.g., Pausas et al. [12]). Therefore, foresters, especially in Germany, often rely on pre-growth of seedlings in tree nurseries and transplanting established containered seedlings into the field sites. This strategy minimizes the need for seed material, but implies additional working steps and costs. To further minimize potential losses, various methods of fast and easy selection of vigorous seedlings have been suggested (for review see e.g., Mattson [13]). For example, in the case of the Mediterranean oak species *Quercus ilex* and *Q. coccifera*, Tsakaldimi et al. [14] tested the use of morphological parameters like height, leaf area or stem diameter as easily measurable parameters for the prediction of seedling survival in the first year after outplanting. Since Mattsson [13] also suggested the use of chlorophyll content and fluorescence as potential, easily measurable predicitive parameters, we attempted to verify their usefulness for the prediction of first-year survival in the field under different climate conditions.

The objectives of the present study were to evaluate the survival and growth responses of seedlings from three oak species established on experimental plantations exposed to different environmental conditions including climate, water availability and soil properties. Specifically, *Quercus ilex* L., *Quercus frainetto* Ten., and *Quercus pubescens* Willd. from Greek and Italian provenances were grown in replicate common gardens in Southern Germany, Central Italy, and Northern Greece. The German plantations also include a local oak species, *Quercus robur* L. for comparative purposes.

The specific questions addressed were:

(1)   Do the three Mediterranean species differ in survival and growth performance during the initial growth phase under different macroclimatic conditions?
(2)   Do Greek and Italian provenances of the Mediterranean species (consistently) show different behavior in different environments?
(3)   Do the local German oaks (*Q. robur*) outperform their Mediterranean relatives in the German plantations?
(4)   Is it possible to predict survival and performance in the field from easily accessible morphological and physiological seedling parameters?

2. Materials and Methods

2.1. Plant Material

In autumn 2015, seeds of three common Mediterranean oak species, *Q. ilex*, *Q. frainetto* and *Q. pubescens* were collected in Central Italy and Northeastern Greece (Supplementary Material Table S1). Seeds from all species were shipped by air mail to commercial tree nurseries in Germany, Italy and Greece, and sown in plastic trays (Quickpot 15T/16 or 24/18). In each country, the trays were filled with different substrate according to the regional practices in order to achieve optimum growth of oaks. Specifically, in Germany trays were filled with a mixture of humus, volcanic lapilli and different organic fibres (FE Typ Bio Blumenerde # 03010, HAWITA Co., Vechta, Germany), in Italy trays were filled with a 40/40/20 mixture of peat, loam and volcanic lapilli along with 25 g·L^{-1} pelleted humified manure and 2 g L^{-1} Osmocote Exact fertilizer, and in Greece trays were filled with a 70/30 mixture of peat and black peat along with 12 kg m^{-3} clay. In Italy and Greece, the emerged seedlings overwintered outdoors, while in Germany they overwintered in a non-heated, but frost-free glasshouse. These procedures are common practice for the first winter with pot-seeded material in the different countries (in Germany, other frost-protection strategies are also applied). In Germany, 1-year-old *Q. robur* seedlings (provenance 81707 "Oberrheingraben", supplied by Darmstädter Forstbaumschulen, Darmstadt, Germany) were added to the ensemble in spring 2016. During 2016, all seedlings were grown outdoors with regular watering, while in Germany seedlings were grown under nettings in autumn. In autumn 2016, emerged seedlings were measured and scored with respect to growth and physiological parameters. In spring 2017, 420 (Germany and Greece) and 210 (Italy) seedlings per species and provenance with the highest score (cf. paragraph "Plant fitness parameter scoring and ranking" below) were transferred to the plantation sites. In each site, 210 seedlings per species and provenance were established, except for Italian *Q. frainetto* in the Greek plantations, where only 170 seedlings were available due to lower seedling emergence rates. In Greece, seedlings were planted on 24 February–4 March 2017, while in Italy and Germany seedlings were planted between 15 April and 1 May 2017.

2.2. Common Garden Experiments

Two plantations were established in Germany, two in Greece, and one in Italy. In Germany, one plantation site is located in a managed oak and pine forest, on fluvial sand, with about 2 m deep groundwater table, in Schwanheim, Frankfurt ("SWA", 50°04′12.6″ N, 08°33′42.2″ E, 114 m a.s.l.). SWA is a typical oak site where oak forests exist for at least 500 years. The other plantation is in Riedberg, Frankfurt ("RIE", 50°10′12.8″ N, 08°37′53.2″ E, 130 m a.s.l.) on loamy soil, where planted seedlings were irrigated when necessary, as defined by monitoring weather conditions and potential drought symptoms on the trees. RIE is a cleared agricultural field, natural vegetation here would be (mixed) beech forest. In Greece, one plantation was established in Olympiada, Chalkidiki ("OLY", 40°36′33.6″ N, 23°45′05.0″ E, 48 m a.s.l.) in natural *Q. ilex* and *Q. pubescens* forest stands with loamy soil, where water was solely provided by precipitation. The particular location of the OLY plantation is a former grazing site. The other plantation is in Stratoniki, Chalkidiki ("STR", 40°31′05.5″ N, 23°47′19.2″ E, 248 m a.s.l.) in a mixed forest in the *Q. pubescens* zone, with loamy soil, where water was also solely provided by precipitation. STR is a cleared forest site. In Italy, the plantation is established at Sant' Anatolia di Narco, Umbria, in the Central Apennine ("SAN", 42°44′18.4″ N, 12°50′17.6″ E, 290 m a.s.l.), surrounded by mixed forest including *Q. ilex* and *Q. pubescens*, on calcareous soil. SAN is a former tree nursery site. During 2017, the seedlings here were also irrigated when necessary. Moreover, weed growth was intensive due to precipitations, thus the sites were cleared 2 to 3 times per year.

Plantation schemes in the common gardens are shown in supplementary Figure S1. In each plantation, 10 plots per species and provenance were established consisting of 21 seedlings, in a 3/5/5/5/3 arrangement called "grouped" scheme (Figure S1D in supplementary material) [15–17]. In the case of Italian *Q. frainetto* in the Greek plantation, only 17 seedlings per plot were established due to the abovementioned limitation of good seedlings. This particular scheme is well suited for enhancing tree quality in oaks, as well as more economic compared to row plantation [15]. To avoid clines in the plantations, the selected trees from the evaluation/scoring procedure were distributed randomly across the plantation in a RCBD design. Except for SAN, the original individual tree IDs from the evaluation procedure in the nursery were recorded for each tree during the planting procedure, thus allowing comparison of evaluation scoring data from 2016 with data measured on trees in later years.

2.3. Climate Data

At RIE and SWA, climate data were recorded on-site with iMetos sm SMT280 weather stations (Pessl Instruments, Weiz, Austria). For SAN, data in Val Nerina, Casteldilago (coordinates 42°34′58.8″ N, 12°46′30″ E at 270 m a.s.l.) were recorded with a Davis Vantage Pro2 weather station (Davis Wetterstationen, Neumarkt, Germany). For OLY and STR, data were recorded at the closest by weather stations of Hellas Gold SA, at 40°36′09.45″ N, 23°45′01.39″ E, 28 m.a.s.l. and at 40°31′03.42″ N, 23°47′47.96″ E, 208 m a.s.l., respectively. Due to technical problems with recording of T_{max} at STR, the T_{max} values from the closest weather station at Arnaia (ARN: 40°29′30.88″ N, 23°36′08.49″ E, 565 m a.s.l.) were used in the STR/ARN diagram in Figure 1. From DOY 146 on in 2017, additional data were collected in the OLY and STR sites with (unshaded) HOBO onset 8K data loggers placed at ca. 1.5 m above ground to address the real heat stress which the seedlings experienced. Precipitation data reported for OLY and STR were all taken from the Hellas Gold weather stations.

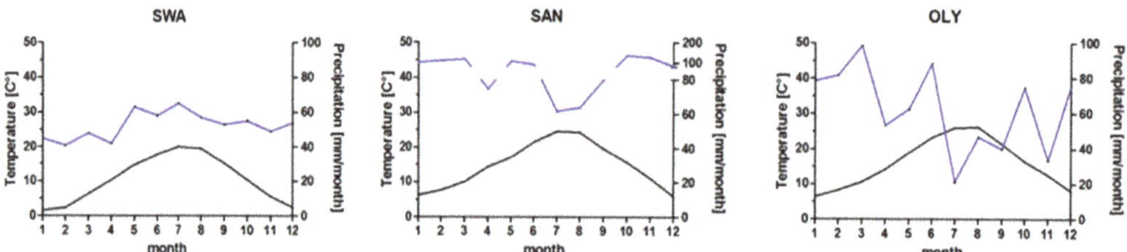

Figure 1. Climate diagrams for SWA, SAN and OLY. Data were averaged for 1981–2010 (SWA: Frankfurt Airport, climate station 1420 of DWD (2018)), 2012–2018 (SAN) and 2008–2017 (OLY), respectively. Black lines represent the average temperature, while blue lines represent the average precipitation.

2.4. Soil Analysis

5–10 representative samples of 200 mL soil each, of the top 20 cm after removal of the top organic layer were randomly collected from various parts of the respective site. Soil pH was measured in 10 mM $CaCl_2$ extracts according to DIN 19 865, part 1 (1977). Granulometric distribution was measured according to DIN 19 683 part 1 and 2 (1973). After rattle sieving with 2 mm sieves, fine soil was subjected to H_2O_2 treatment (for destruction of organic matter) and dispersed in 0.4 N $Na_4P_2O_7$. After wet filtration for particle sizes of coarse (2000–630 μm), medium (630–200 μm) and fine sand (200–63 μm), smaller fractions were analyzed by sedimentation analysis according to Köhn [18]. Inorganic and organic (humus) carbon contents were analyzed according to DIN 19 684, part 2 (1977) with gas analyzers LECO RC-412 and LECO EC-12 (for acidic samples) from LECO Instruments (Mönchengladbach, Germany), respectively. Organic matter was calculated from carbon content of the organic fraction by multiplication with a factor of 1.72 [19]. Initial water contents and field capacity were determined gravimetrically in fresh and water saturated samples, respectively, before and after drying for three days at 105 °C [20].

2.5. Morphological Attributes

In summers of 2016, 2017, 2018 and 2019, seedling heights ("*h*" in the formulas given below) were measured with rulers from soil surface to the top of the longest stem (branch), i.e., in general to the top bud. Root collar diameter was measured with an electronic Vernier caliper at 5 mm above soil level. In 2016, also the numbers of emerged leaves per seedling ("*leaf no*") were counted for fitness determination.

2.6. Relative Chlorophyll Content and Chlorophyll Fluorescence

In general, relative chlorophyll contents were measured in summers of 2016, 2017, 2018 and 2019 with a SPAD 502 Plus chlorophyll meter (Konica Minolta Sensing Co., Munich, Germany) on vital, fully developed leaves emerged in the respective year (important for *Q. ilex*) and given as dimensionless numbers ("*SPAD*" in the formulas below). For comparison with data obtained in Greece in 2016 with an OPTI-Sciences (Hudson, NH, USA) CCM-200 chlorophyll meter, a comparative measuring row of 152 leaves from all oak species covering the observed range of values was performed and used for determination of a calibration coefficient (to obtain "calculated SPAD values" from the CCM-200 values).

Performance index (PI_{abs} in the formulas below) is a parameter summarizing the effects of light trapping, quantum efficiency of reduction of Q_A and efficiency of electron transport from Q_{A-} to the intersystem carriers of the electron chain [21,22]. PI_{abs} was determined in mid- to late summer, i.e., after full leaf development with a Pocket PEA chlorophyll fluorometer (Hansatech, King's Lynn, UK) either at night or in the early morning after dawn, on leaves pre-darkened for a minimum of 20 min with leaf clips, and data were analyzed with PEA Plus 1.0.0.1 (Hansatech) software. A detailed analysis of the results of these analyses is provided in a separate paper [2].

2.7. Plant Fitness Parameter Scoring and Ranking

To compare plant fitness, the morphological and physiological parameters of the emerged seedlings in summer 2016 were used to perform a ranking of the individual seedlings for selection of the best-suited individuals for planting. For this purpose, an arbitrary fitness indicator (*FI*) was calculated for each individual seedling *i* according to (1):

$$FI = 0.25 * \frac{h_i}{h_{max}} + 0.25 * \frac{leaf\ no_i}{leaf\ no_{max}} + 0.25 * \frac{SPAD_i}{SPAD_{max}} + 0.25 * \frac{PIabs_i}{PIabs_{max}} \tag{1}$$

with "i" indicating each individual seedling and "max" indicating the highest value observed for this parameter in the respective population (species/provenance). For the plantations, the 420 (Italy: 210) best performing seedlings were then selected. In Greece, for *Q. frainetto* from Italy only 340 seedlings were available due to low germination rates.

Each individual tree was labeled and could be traced over time in the common garden plantations. To assess the potential predictive value of the 2016 *FI* values for seedling survival and development, we grouped the individual seedlings into "survivors" and "non-survivors" (in 2017) and compared their respective 2016 *FI* values by t-test. Furthermore, within the surviving seedlings, the rank of each individual obtained in 2016 was compared to its rank in 2019 by Spearman's correlation analysis. The 2019 ranking values of the individual were obtained omitting leaf number and SPAD, thus weighing height and PI_{abs} with a factor of 0.5 each. To compare population (species/provenance) performance at the different sites, we took survival rates into account. We calculated, for each site (or, in the case of Germany, for both sites combined, see Table 1), mean values of individual parameters ("x") for each population k (i.e., species/provenance), ("x_k"), then identified the maximum mean values of all population ("x_{max}") and calculated an arbitrary Relative Population Fitness (*RPF*) score (for each site) according to (2):

$$RPF = 0.40 * \frac{survival_k}{survival_{max}} + 0.30 * \frac{h_k}{h_{max}} + 0.30 * \frac{PIabs_k}{PIabs_{max}} \tag{2}$$

Table 1. Relative fitness of populations (*RPF*, cf. Equation (2)) in the different countries and common garden sites. Data were obtained in summer 2019.

Species	Origin	Site	Survival	*RPF*	Ranking
German sites					
Q. pubescens	IT	RIE	98.57	0.869	1
Q. robur	DE	RIE	88.10	0.808	2
Q. pubescens	GR	RIE	96.67	0.804	3
Q. ilex	IT	RIE	95.71	0.777	4
Q. pubescens	GR	SWA	78.57	0.763	5
Q. ilex	GR	RIE	94.76	0.740	6
Q. pubescens	IT	SWA	85.24	0.728	7
Q. robur	DE	SWA	81.90	0.690	8
Q. ilex	IT	SWA	80.48	0.682	9
Q. frainetto	GR	RIE	97.62	0.634	10
Q. ilex	GR	SWA	82.86	0.630	11
Q. frainetto	IT	RIE	95.71	0.623	12
Q. frainetto	IT	SWA	74.29	0.610	13
Q. frainetto	GR	SWA	72.38	0.609	14
Species	**Origin**	**Site**	**Survival**	***RPF***	**Ranking**
Italian site					
Q. pubescens	IT	SAN	81.90	0.944	1
Q. pubescens	GR	SAN	77.62	0.873	2
Q. ilex	IT	SAN	82.38	0.849	3
Q. ilex	GR	SAN	74.76	0.780	4
Q. frainetto	GR	SAN	60.95	0.540	5
Q. frainetto	IT	SAN	58.57	0.486	6
Species	**Origin**	**Site**	**Survival**	***RPF***	**Ranking**
Greek site					
Q. pubescens	GR	OLY	18.10	0.939	1
Q. pubescens	IT	OLY	11.43	0.820	2
Q. ilex	IT	OLY	2.86	0.509	3
Q. frainetto	GR	OLY	3.81	0.443	4
Q. frainetto	IT	OLY	4.29	0.429	5
Q. ilex	GR	OLY	1.90	0.314	6

In addition, the root collar diameter in 2016 was compared (by *t*-test) between the abovementioned "survivors" and "non-survivors" (in 2017) in order to check the parameter's potential as an indicator of second-year survival as reported by Tsakaldimi et al. [14]. Moreover, the potential of seedling height used alone as a predictive parameter for the German and the Greek sites was tested.

2.8. Statistical Analysis

Statistical analysis was conducted using PRISM 2.0 software (GraphPad Software, San Diego, CA, USA). Specifically, Kruskal-Wallis test with Dunn's post-hoc test at a significance level of $p = 0.05$ was conducted to identify potential significant differences between populations (i.e., site/species/provenance combinations).

3. Results

3.1. Soil Parameters

The soil parameters are given in Table 2. The German sites are very different in soil properties, with more calcareous and loamy conditions at RIE and an acidic, sandy soil in the recently partially cleared site SWA with its very high organic matter in the topsoil. The Italian site is typically calcareous and characterized by good water holding capacity. Together with its situation in the river valley, it provides very good conditions for seedling growth. Both Greek sites show slightly acidic, loamy soil with good water holding capacity, but since they both are far away from recent superficial water sources, they both rely on precipitation for seedling water supply.

Table 2. Soil parameters in the plantation sites. Data are means $\pm$ SD of n = 5–6 determinations (except for SWA: means $\pm$ SE, n = 2).

Site	SWA	RIE	SAN	OLY	STR
Granulometric analysis					
Sand (%)	80.8 $\pm$ 0.8	11.6 $\pm$ 4.0	33.6 $\pm$ 6.5	55.2 $\pm$ 5.6	31.4 $\pm$ 4.2
Silt (%)	12.0 $\pm$ 0.3	64.1 $\pm$ 4.3	41.5 $\pm$ 3.7	26.3 $\pm$ 1.9	41.3 $\pm$ 3.5
Clay (%)	7.2 $\pm$ 0.5	24.2 $\pm$ 3.7	24.9 $\pm$ 2.9	18.5 $\pm$ 3.9	27.3 $\pm$ 4.4
Soil type	Sl2	Lu	Ls2-Ls3	Ls4-Sl4	Lt2
pH (CaCl$_2$)	3.3 $\pm$ 0.3	7.2 $\pm$ 0.2	7.2 $\pm$ 0.1	5.4 $\pm$ 0.2	6.3 $\pm$ 0.6
Organic matter (%)	36.5 $\pm$ 19.5	3.1 $\pm$ 2.1	2.2 $\pm$ 0.3	4.7 $\pm$ 1.2	17.6 $\pm$ 1.6
CaCO$_3$ (%)	nd	6.1 $\pm$ 4.6	56.7 $\pm$ 1.8	0	0.1 $\pm$ 0.3
Field capacity (Vol-%)	20.1 $\pm$ 4.6	31.1 $\pm$ 7.6	44.3 $\pm$ 1.1	45.8 $\pm$ 4.0	50.0 $\pm$ 3.8

3.2. Climate Data and Extreme Climate Events

While the young seedlings were under controlled watering conditions during their first year of growth, the trees in SWA, OLY and STR were dependent on precipitation after planting. In RIE and SAN, watering supply was installed and used, if the weather conditions indicated drought situations. Monthly temperature and precipitation patterns at the common gardens (2017, 2018 and 2019) are shown in Table S2, a few of which are noteworthy. In January 2017, the potted seedlings were exposed to extreme frost (subzero temperature down to $-10\,°$C for several days) before establishing the plantations in Greece. In July and August 2018 and 2019, extremely high temperatures in combination with prolonged drought were encountered by the seedlings in Germany. Especially in SWA temperatures were higher than RIE and even exceeded 40 $°$C for a few days.

3.3. Seedling Survival and Growth

Initially, 68% of seeds germinated on average and gave rise to seedlings. The survival rates in the third year after planting (2019) are given in Table 3. Seedlings grown in Germany and Italy exhibited relatively high survival rates, i.e., 59–99%. In Greece, environmental conditions were particularly challenging for the seedlings, especially for *Q. ilex* with few individuals surviving. In general, seedlings in STR had greater survival rates compared to OLY plantation. In all five plantations, *Q. pubescens* revealed the highest survival rates among the three oak species. Moreover, the drier (sandy soil) German plantation site (SWA), and the warmer and drier (lowland) Greek plantation site (OLY) exhibited higher seedling mortality compared to RIE and STR, respectively.

Table 3. Survival rates (%) of seedlings after three years in the five plantation sites (August 2019).

Plantation Site	Species and Provenance					
	Q. Ilex		*Q. Frainetto*		*Q. Pubescens*	
	Greece	Italy	Greece	Italy	Greece	Italy
RIE	95	96	98	96	97	99
SWA	83	80	72	74	79	85
SAN	75	82	61	59	78	82
OLY	2	3	4	5	18	11
STR	6	6	18	18	51	27

Figure 2 depicts the initial size and the height growth of the seedlings during the first three years in the common gardens. As a general pattern, seedlings in Germany were tallest except for the Italian *Q. ilex*, which showed the highest initial size in the Italian nursery (Gubbio, Italy). Statistical analyses after three years in the plantation sites revealed that seedling height (Table S3) was greatest for *Q. ilex* and Italian *Q. pubescens* in RIE and SAN, while *Q. frainetto* seedlings were taller mainly in the German plantations (RIE and SWA) and secondarily in SAN. In the Greek plantations, seedling height of all species was significantly lower showing practically no height growth.

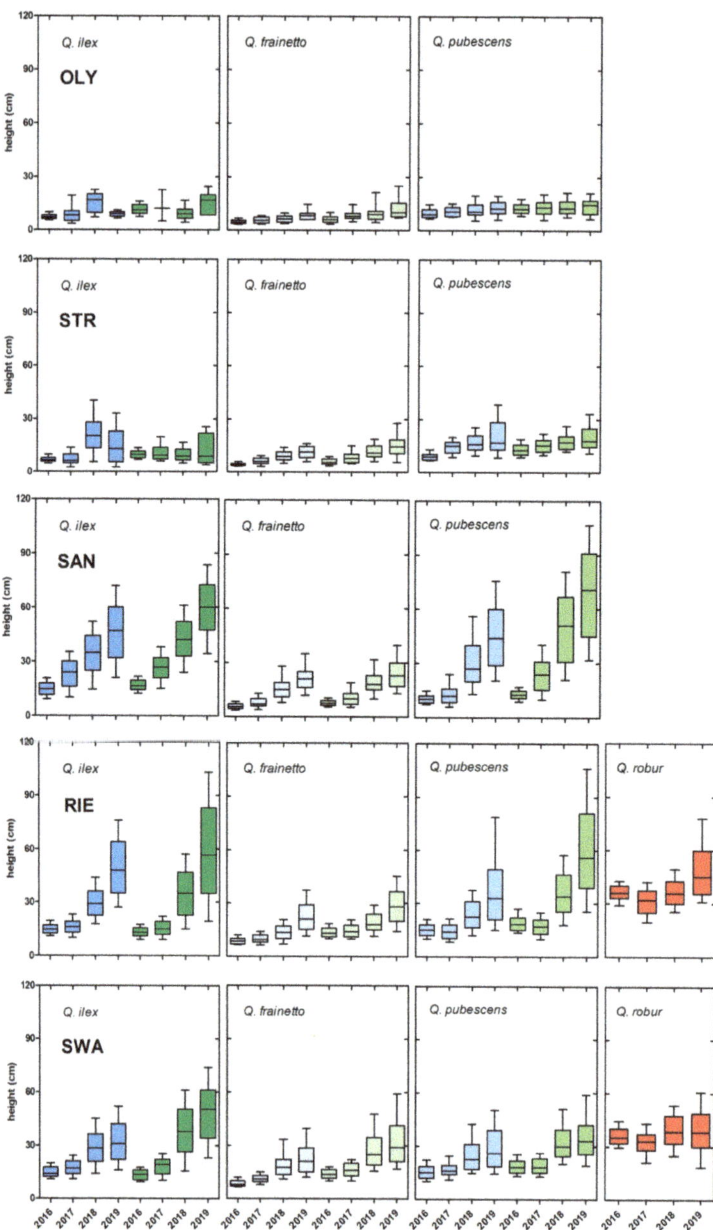

Figure 2. Development of seedling height in the different plantations sites. Greek provenances: blue, Italian provenances: green, *Q. robur*: red. Box plots with 10–90% whiskers.

In general, root collar diameter followed the same trend with height. Height × root collar diameter correlation analysis in 2019 revealed moderately positive (Greek *Q. frainetto* and both *Q. pubescens*) to strong positive (Italian *Q. frainetto* and both *Q. ilex*) relationships between the two parameters (Figure 3).

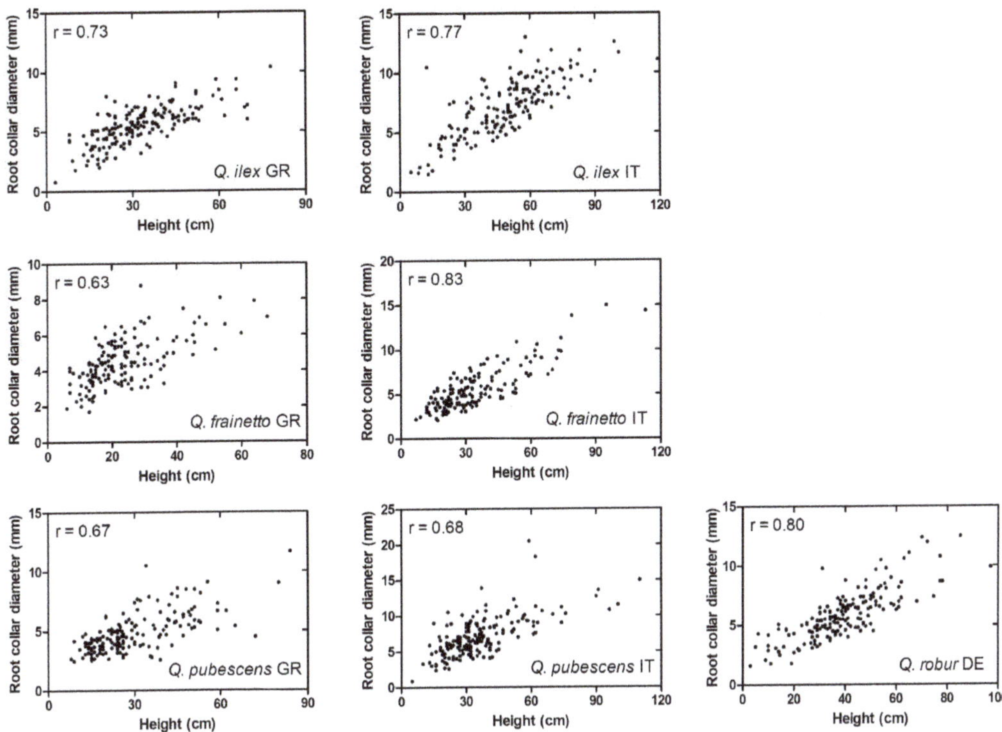

Figure 3. Pearson's correlation analyses between seedling height and root collar diameter in 2019 for every species × provenances tested. The "r" value represents Pearson's correlation coefficient. In all species × provenances, *p*-value was <0.0001.

3.4. Relative Chlorophyll Content

Table 4 depicts the SPAD values of each species established in the common gardens in the nursery and during three years after planting. *Q. ilex* provenances in general showed the lowest SPAD values in all years in STR and high values in SAN and the German sites. SPAD values of both *Q. frainetto* provenances were generally greater in SWA compared to the rest of the plantation sites. Moreover, both Greek and Italian *Q. pubescens* revealed significantly greater SPAD values mainly in SWA and secondarily in SAN. In general, SPAD values in SWA and RIE showed a tendency to increase after establishment in the field, whereas seedlings at SAN revealed the opposite response.

Table 4. Relative chlorophyll content (SPAD) of each species × provenance from the nursery (2016) until after three years in the different plantations sites (2017–2019). Mean values ($\pm$ SE) of the different plantation sites for each species/provenance within a line for each year followed by different letters are significantly different ($P < 5\%$).

Species and Provenance	Year	Plantation Site									
		RIE		SWA		SAN		OLY		STR	
Q. ilex GR	2016	43.49 ± 0.29	a	43.24 ± 0.30	a	33.69 ± 0.55	b	26.10 ± 0.55	c	27.19 ± 0.51	c
	2017	36.83 ± 0.47	b	35.99 ± 0.39	bc	39.74 ± 0.37	a	39.00 ± 1.68	ab	31.70 ± 1.82	c
	2018	36.72 ± 0.43	a	37.33 ± 0.35	a	38.60 ± 0.57	a	30.71 ± 1.03	b	30.64 ± 1.39	b
	2019	33.10 ± 0.33	a	33.84 ± 0.42	a					15.77 ± 5.49	b
Q. ilex IT	2016	43.32 ± 0.33	a	42.88 ± 0.32	a	35.97 ± 0.65	b	33.56 ± 0.54	c	30.36 ± 0.52	d
	2017	35.49 ± 0.44	b	36.30 ± 0.40	b	41.18 ± 0.41	a	34.73 ± 3.89	ab	34.24 ± 1.97	b
	2018	39.14 ± 0.49	a	38.59 ± 0.30	a	40.19 ± 0.52	a	30.97 ± 2.45	b	29.06 ± 1.65	b
	2019	33.95 ± 0.41	a	34.81 ± 0.38	a					22.13 ± 4.40	b
Q. frainetto GR	2016	38.76 ± 0.31	a	38.34 ± 0.33	a	18.09 ± 0.29	c	24.72 ± 0.63	b	25.22 ± 0.63	b
	2017	33.66 ± 0.47	b	37.10 ± 0.49	a	30.28 ± 0.56	c	30.98 ± 1.28	bc	30.05 ± 0.96	c
	2018	33.51 ± 0.38	b	37.08 ± 0.31	a	24.48 ± 0.89	d	23.75 ± 1.06	d	28.08 ± 0.80	c
	2019	27.51 ± 0.51	b	35.74 ± 0.34	a					19.93 ± 1.01	c
Q. frainetto IT	2016	40.84 ± 0.27	a	41.42 ± 0.26	a	19.47 ± 0.49	c	23.52 ± 0.59	b	23.82 ± 0.60	b
	2017	32.01 ± 0.42	b	36.61 ± 0.45	a	32.54 ± 0.48	b	30.88 ± 1.02	b	32.71 ± 1.06	b
	2018	30.85 ± 0.41	b	36.44 ± 0.32	a	24.48 ± 0.82	c	22.23 ± 0.88	c	31.06 ± 1.19	b
	2019	25.97 ± 0.48	b	34.72 ± 0.34	a					16.70 ± 1.47	c
Q. pubescens GR	2016	41.85 ± 0.30	a	41.93 ± 0.30	a	29.29 ± 0.54	c	33.65 ± 0.55	b	35.25 ± 0.53	b
	2017	37.05 ± 0.40	b	40.16 ± 0.36	a	40.45 ± 0.39	a	33.56 ± 0.60	c	34.54 ± 0.69	c
	2018	37.79 ± 0.32	b	40.18 ± 0.23	a	39.15 ± 0.49	ab	28.90 ± 0.76	c	31.64 ± 0.59	c
	2019	36.23 ± 0.28	b	38.48 ± 0.25	a					29.07 ± 1.32	c
Q. pubescens IT	2016	40.57 ± 0.24	a	40.76 ± 0.26	a	27.25 ± 0.46	b	28.52 ± 0.51	b	27.37 ± 0.50	b
	2017	36.08 ± 0.41	b	38.15 ± 0.39	a	39.20 ± 0.40	a	32.99 ± 0.88	c	32.19 ± 0.83	c
	2018	35.19 ± 0.32	c	39.66 ± 0.25	a	37.50 ± 0.47	b	23.59 ± 0.84	e	28.43 ± 0.68	d
	2019	34.77 ± 0.32	b	37.53 ± 0.36	a					19.83 ± 1.16	c
Q. robur DE	2016	37.54 ± 0.41	a	37.89 ± 0.36	a						
	2017	27.73 ± 0.51	b	36.14 ± 0.57	a						
	2018	33.25 ± 0.46	b	36.10 ± 0.42	a						
	2019	32.39 ± 0.45	b	37.23 ± 0.44	a						

3.5. Quality Scoring of Seedlings as Predictors for Survival and Growth and Population Fitness

To analyze, whether root collar diameter measured in the first year to predict second-year survival of oak seedlings, the length data from 2016 of the perished (2017) and the live seedlings in 2017 were arranged into separate groups for all seedlings planted in Germany (favourable conditions) and in Greece (unfavourable conditions), respectively, and yielded statistically significant differences were tested, as suggested by Tsakaldimi et al. [14]. No significant differences were observed between survived (2017) and perished seedlings in SWA and STR. However, survived *Q. frainetto* from both provenances and *Q. pubescens* IT in RIE, and *Q. frainetto* IT in OLY developed significantly thicker stems (in 2016) compared to the respective perished seedlings (Supplemental Material Table S4). To further elucidate potential differences between the survivors and the perished trees at the sites, where stem diameter alone revealed no indicative value, a more complex parameter was also developed, taking into account both morphological and physiological fitness (i.e., *FI*, a function of tree length, leaf development, chlorophyll content and efficiency of the photosynthetic apparatus (Equation (1)). However, in no combination of provenance and country of plantation the 2016 *FI* values of the subpopulations (live in 2017 versus dead in 2017) yielded significant differences, meaning that the 2016 *FI* values could not be used as predictors for seedling survival (Supplemental Material Table S5).

In addition, the relationship between the *FI* values of individual trees in 2016 and 2019, respectively, was studied, for the German plantations to assess the predictive value of seedling measurements for future growth (Supplemental Material Table S6). In the German plantations, where 72–99% of the seedlings survived into 2019 in all provenances, the Spearman correlation coefficients between the *RPF* in 2016 and 2019 were weak (i.e., <0.21) and calculated *RPF* obtained in 2016 were considered of little predictive value for the future performance of individual seedlings. If data for seedling height in 2016 were compared to those of 2019, in all cases weak correlations were found both in the German and in the Greek plantations (Pearson's r < 0.3).

In Table 1, all populations per site have been evaluated for their relative vitality, using survival rates, morphometric and physiological data obtained in 2019. Data for STR had to be omitted due to road blocking by storm-felled trees in 2019 which prevented accessibility at night for PI_{abs} measurements. However, the site was accessible at daytime and data for survival and height at STR are given in Table 3 and Figure 2, respectively. The best performing population in Germany was the Italian *Q. pubescens* followed by *Q. robur* and the Greek *Q. pubescens*, all in RIE, while *Q. frainetto* showed the lowest *RPF*. In Germany, RIE populations showed higher vitality scores compared to SWA. In Italy, *Q. pubescens* showed the best overall performance with *Q. ilex* following. In Greece, *Q. pubescens* showed relatively higher vitality scores compared to the other oak species.

4. Discussion and Conclusions

The present article provides an early evaluation of three oak species' establishment in four common garden studies, including nursery stage as well as field performance during the first three years after field plantation. The seedling production process followed the protocols of local nurseries. It was decided to not impose a standardized procedure on the nurseries, since each of them had its own year-long experience with the local oak provenances under the local climatic conditions. The same argument holds for the different plantation procedures. It was not the aim of the study to experiment with the local standardized growth techniques, since this would have introduced additional technical problems in procedure (e.g., shipping of plantation soil from one country to another, missing availability of frost-free greenhouses in Gubbio and Olympiada) and site management. The authors were aware that the different treatments at different sites would prevent a more comprehensive comparison of seedling performance with solely the different macroclimates (and soil conditions) as variables. However, the cooperation with the local nurseries and their local experiences with the growth of oaks reflect the situation in real-life transfer of genotypes from one country to the other.

The potential to include a German population of *Q. pubescens* was also examined. In German nurseries, no *Q. pubescens* seeds or seedlings are available from German populations, since these are mostly very small, often include hybrids with the local species, and, to our knowledge, are thus usually not included in certified seedling material lists. It was decided against self-collecting such seeds due to the genetic uncertainties as well as due to the fact that it was not possible to predict if there would be sufficient seeds available in 2016. Since the aim of the study was not translocating German material to the Mediterranean countries, the German control material was confined to the local species at the plantation sites (i.e., *Q. robur*).

4.1. Seedling Survival and Establishment

Seedling emergence was good for all collected seed populations, giving rise to sufficient plant material for an initial selection step, except for *Q. frainetto* of Italian origin at the Greek nursery. Transportation of the seeds in perforated plastic bags by airmail may have led to partial anoxia in the center of the bags, thus reducing germination rates, but in general, the outcome of seedlings was satisfactory. After their respective plantation in winter 2016/17, survival rates were assessed in summer 2019, and except for the Greek plantations, seedlings had well established themselves in the Common Gardens (Table 3). Initially it was intended to irrigate the seedlings in one of the Greek sites but this ultimately turned out to be impossible due to technical problems, i.e., the large distance in the finally available plantation site from available water. At RIE and SAN, facilities for watering of the seedlings had been established beforehand, so during severe soil drought in summer, these seedlings could be watered. Here, as well as in SWA, where no watering occurred, survival rates of all seedlings were satisfactory. In Greece, seedlings encountered a period of extreme frost in early January 2017 before establishing the plantations, and the evergreen *Q. ilex* was severely affected. Low seedling vitality has also been described for other broadleaves after freezing [23] and in summer-dry, calcareous Mediterranean habitats (e.g., Pausas et al. [12]: only ca. 15% survival of *Q. ilex* ssp. *ballota* after 3 years). During planting at the beginning of March, the seedlings still looked alive with most of the leaves still green, but turned out to be lethally damaged when the growth period started. For *Q. ilex* from its northernmost population in Italy, North of Lake Garda, it is known that frost tolerance (50% survival) of 1-year-old seedlings is around −10 (leaves) to −15 °C (sprout cambium) [24]. Such temperatures have, to our knowledge, never been recorded in the area of Olympiada, where the seedling material stems from, nor in Latium, Italy, close to the sea, where the Italian seeds have been collected. Since temperatures at soil level, where the seedlings overwintered outdoors, may also have been well below the −9.7 °C recorded at the Olympiada weather station (i.e., at 2 m height), we therefore conclude, that stems, roots and leaves of most *Q. ilex* seedlings were irreversibly damaged already before the planting. In contrast to the seedlings, the predominant mature *Q. ilex/Q. pubescens* forest in the surroundings of the plantation at OLY was not visibly affected by the frost event (unpublished field observations in May 2017), in accordance with the findings at Lake Garda, where adult trees revealed frost tolerance of −25 °C (stem cambium, 50% survival, [24]). The deciduous species, *Q. frainetto* and *Q. pubescens*, which occur naturally at higher elevations in Chalkidiki, northern Greece, obviously developed better frost tolerance of stem and cambium (and even roots). Although *Q. frainetto* seedling survival rates in the field are better when planted earlier and allowed to develop a better root system (i.e., Dec/Jan, [25]), in the current experiment we opted for a later planting date because of the severe frost in January 2017. However, in March/early April 2017 precipitation was low, especially at OLY (Table S2), and many seedlings died of drought due to weakly developed root systems before sufficient rainfalls occurred in mid-April. *Q. pubescens* showed better survival rates compared to *Q. frainetto* under identical conditions (Table 1) confirming similar findings in Northern Greece [26]. *Q. frainetto* has been found to be the least drought-tolerant oak among four Mediterranean species including *Q. pubescens* and *Q. ilex* [27], and this may explain its low survival rate at OLY and STR after the spring drought immediately after plantation. At SAN and RIE, in contrast, seedlings were watered when necessary and therefore mortality was much lower. In SWA, although seedlings relied on precipitation, mortality was also low. Here, strong rainfalls occurred in mid-April 2017 before the flushing of the trees, which is usually observed one month later in Germany than in Greece.

Concerning the predictive value of morphological and physiological properties (chl content, PI_{abs}) of 1-year old seedlings for their future survival and development in the field, our results did not meet our expectations. Bayala et al. [28], on five tropical tree species, and Tsakaldimi et al. [14] on five Mediterranean tree species, including *Q. ilex* and *Q. coccifera*, pointed to stem diameter as a good indicator for second-year survival after outplanting. We examined this proposed indicator and the results were variable among the common gardens. In RIE, where seedlings showed the greatest survival, two-year old seedlings of *Q. frainetto* (both provenances) and *Q. pubescens* IT were successfully grouped to survived or perished depending on their root collar diameter, thus confirming the results of Tsakaldimi et al. [14] (Table S4). In general, root collar diameter of *Q. frainetto* showed a greater potential to predict second-year survival but differences were not always significant. While the morphological features height and root collar diameter were well correlated in older seedlings, confirming data of Pausas et al. [12] on *Q. ilex* ssp. *ballota* (Figure 2), the *FI* values in the subgroups of established (survivors into 2017) and perished seedlings (dead in 2017) did not differ significantly (Table S5). Secondly, the Spearman r values between the *FI* values in 2016 (using seedling height, number of leaves, chl content and PI_{abs}) and 2019 (using seedling height and PI_{abs}) were considered too low to be of practical value (cf. Supplementary Material Table S6: r was either not significant or even negative in all but one cases). Quite similarly, if measurements are confined to height the predictive value is also weak, with Pearson coefficients for the respective 2016 and 2019 values being negative or below 0.17 in all cases, thus height alone is neither considered suitable for preselection for plantation purposes.

In conclusion, of the parameters tested, only in good growing conditions (i.e., RIE), root collar diameter of 2nd year seedlings had a moderate predictive value for seedling survival in three out of seven species/provenances, while other non-destructive parameters (included in the *FI* values) appear to be of no practical value.

4.2. Cumulative Growth Conditions at the Common Garden Sites as Reflected in Seedling Performance

Height and root collar growth during the first three years after planting (measured in summers 2017, 2018 and 2019) can be used as first predictors for the suitability of the respective oak provenances for establishment at the different sites. Generally, both parameters were relatively high at SAN for most species/provenances over the three years, reflecting the good nutrient conditions (cations) in the basic soil and the sufficient water supply. In the German sites, *Q. pubescens* and *Q. robur* revealed no height growth during the first year, pointing to the costs of acclimation (i.e., root development) after planting. However, in the following years the Italian *Q. pubescens* in RIE developed strongly and reached comparable height with the SAN homologue most probably due to a well-developed root system. In the first year at the German sites, *Q. frainetto* and *Q. ilex* seedlings at SWA in general showed a stronger height growth than at RIE, possibly because of the lower light availability at SWA (thinned pine umbrella plantation with 50 trees/ha [5]) as compared to the open setup at RIE. However, after three years both *Q. ilex* provenances as well as *Q. robur* were significantly promoted by the relatively favorable conditions in RIE (i.e., irrigation) compared to SWA. After three years at the Greek sites, especially *Q. pubescens* revealed better height growth at STR than at OLY, again possibly due to partial shading (cf. Supplementary Material Table S3).

The quite consistent higher chl contents of *Q. frainetto*, *Q. pubescens* and *Q. robur* at SWA vs. RIE can be explained by partial shading by pine trees at SWA, whereas seedlings at RIE grew in full sunlight. At SAN, seedlings were also partially shaded by weeds during the period of measurements which caused early senescence of *Q. frainetto* and subsequently low chl content values. Leaf senescence depends on the carbon fixation efficiency which is modulated by plant photoreceptors such as phytochrome A [29]. STR and OLY seedlings generally showed similar (low) chl content over the three-year period as compared to the other sites. At STR, the high irradiation, at least compared to SWA and RIE, may have caused the lower chl contents. Additionally, OLY seedlings were exposed to full sunlight, combined with high temperature and a long drought period. This combination accelerated photoinhibition [2]. Frequent photoinhibition, in turn, may cause photodamage to pigments. Similarly, Cotrozzi et al. [30] reported leaf injuries (i.e., yellowing) induced by drought stress in seedlings of *Q. ilex*, *Q. pubescens* and *Q. cerris*. In the German plantations, lower SPAD values were observed in 2017, 2018 and 2019 (outdoor grown seedlings) vs. data from 2016 (greenhouse-grown seedlings).

The advantage which the seedlings in Germany had by initial greenhouse growth over the Italian seedlings at the time of planting (compare seedling heights in 2016 in Figure 2) was (mostly) lost in the course of outdoor growth in 2019. This is partly due to prolonged extreme heat and drought periods in 2018 and 2019 (Table S2) which showed similar intensity with the heat and drought in Europe in 2003 [31]. The most severe growth conditions existed at the Greek sites, nearly extinguishing the *Q. ilex* populations by the very unusual extreme frost in January 2017 and affecting growth in the deciduous species e.g., by spring drought.

4.3. Preliminary Conclusions for (Re)forestation Strategies

In summary, we can conclude that:

(1) Of the Mediterranean species, *Q. pubescens* shows the best performance at all sites, followed by *Q. ilex*. *Q. frainetto* showed the lowest performance under all conditions, except for Greek *Q. ilex* at OLY. Since this finding was independent of provenance and site, it appears to be an inherent trait and we conclude that among the deciduous species, *Q. frainetto* is a "slow-growing" species (sensu Lambers and Poorter, [32]) as opposite to *Q. pubescens* (and *Q. robur*).
(2) Whether Greek or Italian provenances perform better at a given site depends on site conditions. In general, the differences between the provenances were small.
(3) In Germany, *Q. robur* did not outperform *Q. pubescens*. Both sites (RIE and SWA) were quite warm and suffered from heatwaves during the experimental period, which may have masked a potential advantage (under milder temperatures and higher precipitation) of the local species over the Mediterranean one.
(4) The predictive value of the morphological and physiological measurements on the seedlings in 2016 is not considered of practical use, except for root collar diameter in certain combinations of provenance and growth site.

Presently, differences between Italian and Greek populations are small, and it will be necessary to monitor seedling development for more years to decide, whether the Italian populations show better performance in the long run. At OLY and STR, growth conditions are extremely hard for young oaks. To regenerate *Q. ilex* forest (especially at the lower, dryer site OLY), several attempts in successive years may be necessary to avoid frost damage, if extreme situations like in January 2017 are to occur in future years again. It is further recommended to irrigate new plantations of rooted seedlings in this region during the first one or two years [33]. In Germany, *Q. pubescens* and *Q. robur* showed the highest relative vitality scores (Table 1) in both sites, despite the fact that the two species showed minimal height growth during the first year after planting. *Q. frainetto* showed inferior quality compared to the other oaks. Consequently, *Q. pubescens* may be considered a potential future forest tree on poor and dry soils for the expected future summer heat and drought conditions in Central Europe, as were observed in Germany in 2018 and 2019. Whether Italian or Greek provenances are better suited for planting of Mediterranean species in southwest Germany should be addressed by future research in the common gardens.

Supplementary Materials: The following are available online at https://www.mdpi.com/article/10.3390/f12060678/s1, TableS1: Oak species, collection sites and nursery sites. Table S2: Average minimum and maximum monthly temperatures and total monthly precipitation recorded in the common gardens in the course of 2017, 2018 and 2019. Data in RIE, SWA were recorded on site, data in SAN at the climate station in Casteldilago, and data for OLY and STR close to the sites at the climate stations of Hellas Gold. Table S3: Statistical analysis of height (in cm) in the common gardens, measured in summer 2019. Values followed by different letters within a row are significantly different at $P < 5\%$, as calculated by Kruskal-Wallis one-way-ANOVA tests with Dunn´s post-hoc test. Means of *Q. robur* were analyzed by t-test at $P < 5\%$. Table S4: Statistical analysis of root collar diameter (mm) in the common gardens, measured in summer 2016, to examine the parameters potential as indicator of second-year survival after outplanting. Values followed by different letters within a row are significantly different at $P < 5\%$, as calculated by t-tests. Table S5: Mean values of fitness indicators (*FI*) of the plants in 2016, separated into those who established live in 2017 and those who perished in the first year after plantation (dead in 2017). Table S6: Spearman correlation analyses between Relative Population Fitness (*RPF*) ranking positions in 2019 vs. 2016 at the German sites. *RPF* were calculated as described in Materials and methods. Pearson´s r value was also calculated for the height data in 2019 vs. 2016 at the German and Greek sites. Significance values for the correlations are given at $P < 5\%$ (*), 1% (**), $0,1\%$ (***). Figure S1: Plantation schemes at RIE, SWA, and SAN (1A), OLY (1B), and STR (1C). Within each plot, individual trees were planted according to 1D; in the case of *Q. frainetto* from Italy at OLY and STR, trees number 02, 09, 13 and 20 were omitted.

Author Contributions: Conceptualization, W.B.; methodology, W.B., K.R. and F.B. (Filippo Bussotti); investigation, F.B. (Filippos Bantis), J.G. and E.F.; data curation, W.B., E.F. and F.B. (Filippos Bantis); writing—original draft preparation, F.B. (Filippos Bantis) and W.B.; writing—review and editing, F.B. (Filippos Bantis), J.G., E.F., K.R., F.B. (Filippo Bussotti) and W.B.; supervision, project administration and funding acquisition, W.B. All authors have read and agreed to the published version of the manuscript.

Funding: This work was financed by grant No. 01DS15014 by the German Federal Ministry of Education and Research (BMBF) to W.B.

Data Availability Statement: The data presented in this study are available within the article and its supplementary materials.

Acknowledgments: Special thanks are due to the City of Frankfurt/Main (Germany), to Greek Nurseries SA (Olympiada, Greece), Hellas Gold SA (Athens, Greece) and to Agenzia Forestale Regionale (Perugia, Umbria, Italy) for providing the respective plantation sites and help with establishment and maintenance of the plantations as well as for providing climate data (OLY and STR). Thanks are due to Martina Pollastrini (Univ. Firenze), Vera Holland, Sonja Ströll, Nathalie Reininger and Lisa Schäfer (Univ. Frankfurt) for help with the data acquisition, and also to Mariangela Fotelli for commenting on the manuscript.

Conflicts of Interest: The authors declare no conflict of interest.

References

1. Frank, D.; Reichstein, M.; Bahn, M.; Thonicke, K.; Frank, D.; Mahecha, M.D.; Smith, P.; van der Velde, M.; Vicca, S.; Babst, F.; et al. Effects of climate extremes on the terrestrial carbon cycle: Concepts, processes and potential future impacts. *Glob. Chang. Biol.* **2015**, *21*, 2861–2880. [CrossRef]
2. Bantis, F.; Fruchtenicht, E.; Graap, J.; Stroll, S.; Reininger, N.; Schafer, L.; Pollastrini, M.; Holland, V.; Bussotti, F.; Radoglou, K.; et al. The JIP-test as a tool for forestry in times of climate change. *Photosynthetica* **2020**, *58*, 224–236. [CrossRef]
3. Hanewinkel, M.; Cullmann, D.A.; Schelhaas, M.J.; Nabuurs, G.J.; Zimmermann, N.E. Climate change may cause severe loss in the economic value of European forest land. *Nat. Clim. Chang.* **2013**, *3*, 203–207. [CrossRef]
4. Bussotti, F.; Pollastrini, M.; Holland, V.; Brüggemann, W. Functional traits and adaptive capacity of European forests to climate change. *Environ. Exp. Bot.* **2015**, *111*, 91–113. [CrossRef]
5. Koller, S.; Jedmowski, C.; Kamm, K.; Brüggemann, W. The South Hesse Oak Project (SHOP): Species- and site-specific efficiency of the photosynthetic apparatus of Mediterranean and Central European Oaks in Central Europe. *Plant Biosyst.* **2014**, *148*, 237–248. [CrossRef]
6. Mátyás, C. Climatic adaptation of trees: Rediscovering provenance tests. *Euphytica* **1996**, *92*, 45–54. [CrossRef]
7. Nicotra, A.B.; Atkin, O.K.; Bonser, S.P.; Davidson, A.M.; Finnegan, E.J.; Mathesius, U.; Poot, P.; Purugganan, M.D.; Richards, C.L.; Valladares, F.; et al. Plant phenotypic plasticity in a changing climate. *Trends Plant Sci.* **2010**, *15*, 684–692. [CrossRef] [PubMed]

8. Vitasse, Y.; Delzon, S.; Bresson, C.C. Altitudinal differentiation in growth and phenology among populations of temperate-zone tree species growing in a common garden. *Can. J. For. Res.* **2009**, *39*, 1259–1269. [CrossRef]
9. Williams, M.R.; Dumroese, R.K. Preparing for climate change: Forestry and assisted migration. *J. For.* **2013**, *111*, 287–297. [CrossRef]
10. Koralewski, T.E.; Wang, H.H.; Grant, W.E.; Byram, T.D. Plants on the move: Assisted migration of forest trees in the face of climate change. *For. Ecol. Manang.* **2015**, *344*, 30–37. [CrossRef]
11. Zeng, X.; Fischer, G.A. Using multiple seedlots in restoration planting enhances genetic diversity compared to natural regeneration in fragmented tropical forests. *For. Ecol. Manag.* **2021**, *482*, 118819. [CrossRef]
12. Pausas, J.G.; Bladé, C.; Valdecantos, A.; Seva, J.P.; Fuentes, D.; Alloza, J.A.; Vilagrosa, A.; Bautista, S.; Cortina, J.; Vallejo, R. Pines and oaks in the restoration of Mediterranean landscapes of Spain: New perspectives for an old practice—A review. *Plant Ecol.* **2004**, *171*, 209–220. [CrossRef]
13. Mattsson, A. Predicting field performance using seedling quality assessment. *New For.* **1996**, *13*, 223–248.
14. Tsakaldimi, M.; Ganatsas, P.; Jacobs, D.F. Prediction of planted seedling survival of five Mediterranean species based on initial seedling morphology. *New For.* **2013**, *44*, 327–339. [CrossRef]
15. Saha, S.; Kuehne, C.; Kohnle, U.; Brang, P.; Ehring, A.; Geisel, J.; Leder, B.; Muth, M.; Petersen, R.; Peter, J.; et al. Growth and quality of young oaks (*Quercus robur* and *Quercus petraea*) grown in cluster plantings in central Europe: A weighted meta-analysis. *For. Ecol. Manang.* **2012**, *283*, 106–118. [CrossRef]
16. Saha, S.; Kuehne, C.; Bauhus, J. Lessons learned from oak cluster planting trials in central Europe. *Can. J. For. Res.* **2016**, *47*, 139–148. [CrossRef]
17. Skiadaresis, G.; Saha, S.; Bauhus, J. Oak group planting produces a higher number of future crop trees, with better spatial distribution than row planting. *Forests* **2016**, *7*, 289. [CrossRef]
18. Köhn, M. Die mechanische Analyse des Bodens mittels Pipettmethode. *J. Plant Nutr. Soil Sci.* **1931**, *21*, 211–222. [CrossRef]
19. Ad.Hoc-AG Boden. Bodenkundliche Kartieranleitung KA5 (Manual of Soil Mapping. In *Bundesanstalt für Geowissenschaften und Rohstoffe*, 5th ed.; Eckelmann, W., Ed.; Schweizerbart: Frankfurt, Germany, 2005.
20. Alfonso, S. Einfluss von Trockenstress auf die Photosynthese in der Gattung. Ph.D. Thesis, University of Frankfurt, Frankfurt, Germany, 2010. (In German)
21. Strasser, R.J.; Srivastava, A.; Tsimilli-Michael, M. The fluorescence transient as a tool to characterize and screen photosynthetic samples. In *Probing Photosynthesis: Mechanisms, Regulation and Adaption*; Yunus, M., Pathre, U., Mohanty, P., Eds.; Taylor & Francis: London, UK, 2000; pp. 445–483.
22. Strasser, R.J.; Tsimilli-Michael, M.; Srivastava, A. Analysis of the fluorescence transient. In *Chlorophyll a Fluorescence: A Signature of Photosynthesis*; Papageorgiou, G.C., Govindjee, Eds.; Springer: Dordrecht, The Netherlands, 2004; pp. 321–362.
23. Radoglou, K.; Raftoyannis, Y. Effects of desiccation and freezing on vitality and field performance of broadleaved tree species. *Ann. For. Sci.* **2001**, *58*, 59–68. [CrossRef]
24. Larcher, W. Zunahme des Frostabhärtungsvermögens von *Quercus ilex* im Laufe der Individualentwicklung. *Planta* **1969**, *88*, 130–135. [CrossRef] [PubMed]
25. Radoglou, K.; Raftoyannis, Y.; Halivopoulos, G. The effect of planting date and seedlings quality on field performance of Castanea sativa Mill and Quercus frainetto Ten. *Forestry* **2003**, *76*, 569–578. [CrossRef]
26. Siam, A.M.J.; Radoglou, K.; Noitsakis, B.; Smiris, P. Ecophysiology of seedlings of three deciduous oak trees during summer water deficit. *Sudan J. Des. Res.* **2009**, *1*, 71–87.
27. Fotelli, M.N.; Radoglou, K.M.; Constantinidou, H.I.A. Water stress responses of seedlings of four Mediterranean oak species. *Tree Physiol.* **2000**, *20*, 1065–1075. [CrossRef] [PubMed]
28. Bayala, J.; Dianda, M.; Wilson, J.; Ouedraogo, S.J.; Sanon, K. Predicting field performance of five irrigated tree species using seedling quality assessment in Burkina Faso, West Africa. *New For.* **2009**, *38*, 309–322. [CrossRef]
29. Brouwer, B.; Ziolkowska, A.; Bagard, M.; Keech, O.; Gardestrom, P. The impact of light intensity on shade-induced leaf senescence. *Plant Cell Environ.* **2012**, *35*, 1084–1098. [CrossRef] [PubMed]
30. Cotrozzi, L.; Remorini, D.; Pellegrini, E.; Landi, M.; Massai, R.; Nali, C.; Guidi, L.; Lorenzini, G. Variations in physiological and biochemical traits of oak seedlings grown under drought and ozone stress. *Physiol. Plant* **2016**, *157*, 69–84. [CrossRef]
31. Rebetez, M.; Mayer, H.; Dupont, O.; Schindler, D.; Gartner, K.; Kropp, J.P.; Menzel, A. Heat and drought 2003 in Europe: A climate synthesis. *Ann. For. Sci.* **2006**, *63*, 569–577. [CrossRef]
32. Lambers, H.; Poorter, H. Inherent variation in growth rate between higher plants: A search for physiological causes and ecological consequences. *Adv. Ecol. Res.* **1992**, *23*, 187–261.
33. David, T.S.; Pinto, C.A.; Nadezhdina, N.; David, J.S. Water and forests in the Mediterranean hot climate zone: A review based on a hydraulic interpretation of tree functioning. *For. Syst.* **2016**, *25*, eR02. [CrossRef]

Article

Climate Change Effects in a Mediterranean Forest Following 21 Consecutive Years of Experimental Drought

Romà Ogaya [1,2,*] and Josep Peñuelas [1,2]

[1] CSIC, Global Ecology Unit CREAF-CSIC-UAB, 08193 Bellaterra, Catalonia, Spain; Josep.Penuelas@uab.cat
[2] CREAF, 08193 Cerdanyola del Vallès, Catalonia, Spain
* Correspondence: r.ogaya@creaf.uab.cat

Abstract: *Research Highlights:* A small, long-term decrease in the water availability in a Mediterranean holm oak forest elicited strong effects on tree stem growth, mortality, and species composition, which led to changes in the ecosystem function and service provision. *Background and Objectives:* Many forest ecosystems are increasingly challenged by stress conditions under climate change. These new environmental constraints may drive changes in species distribution and ecosystem function. *Materials and Methods:* An evergreen Mediterranean holm oak (*Quercus ilex* L.) forest was subjected to 21 consecutive years of experimental drought (performing 30% of rainfall exclusion resulted in a 15% decrease in soil moisture). The effects of the annual climatic conditions and the experimental drought on a tree and shrub basal area increment were studied, with a focus on the two most dominant species (*Q. ilex* and the tall shrub *Phillyrea latifolia* L.). *Results:* Stem growth decreased and tree mortality increased under the experimental drought conditions and in hot and dry years. These effects differed between the two dominant species: the basal area of *Q. ilex* (the current, supradominant species) was dependent on water availability and climatic conditions, whereas *P. latifolia* was more tolerant to drought and experienced increased growth rates in plots where *Q. ilex* decay rates were high. *Conclusions:* Our findings reveal that small changes in water availability drive changes in species growth, composition, and distribution, as demonstrated by the continuous and ongoing replacement of the current supradominant *Q. ilex* by the subdominant *P. latifolia*, which is better adapted to tolerate hot and dry environments. The consequences of these ecological transformations for ecosystem function and service provision to human society are discussed.

Keywords: basal area increment; climate change; forest dieback; Mediterranean forest; stem growth; water availability

Citation: Ogaya, R.; Peñuelas, J. Climate Change Effects in a Mediterranean Forest Following 21 Consecutive Years of Experimental Drought. *Forests* **2021**, *12*, 306. https://doi.org/10.3390/f12030306

Academic Editor: Mariangela Fotelli

Received: 14 January 2021
Accepted: 1 March 2021
Published: 6 March 2021

1. Introduction

Climate change is increasing the levels of environmental stress conditions for many forest ecosystems around the world due to higher rates of evapotranspiration and increases in the intensity and duration of extreme climate events [1,2]. Some water-limited forest ecosystems, such as Mediterranean forests, may be particularly vulnerable to the expected decrease in water availability and may experience associated decreases in stem growth [3,4], greater levels of canopy defoliation [5], and increases in tree mortality [6,7]. While rising concentrations of atmospheric carbon dioxide (CO_2) are associated with higher growth rates and productivity in forests [8], it is possible that decreases in stem growth may occur in water-limited ecosystems, if the effects of water scarcity exceed those of fertilization from the increasing levels of CO_2.

Quercus ilex L. is an evergreen tree species and it is the most abundant and representative of the Mediterranean forest [9]. *Phillyrea latifolia* L. is a tall shrub evergreen species, and it is also very common in the Mediterranean forest. Both species are well adapted to resist a drought period (usually the summer season) due to their stomatal closure around midday to avoid excessive water loss by transpiration [9]. Despite these two species having

39

experienced similar distributions and adaptations for coping with drought conditions, several studies have shown that *Q. ilex* is more sensitive to dry and hot conditions than other co-occurring tall shrub species, such as *P. latifolia* [3,4,10,11]. For example: within a shared forest, *Q. ilex* was found to suffer from high rates of stem mortality during periods of extreme drought, which had little impact on *P. latifolia* [10,12]; while wood carbon allocation decreased in *Q. ilex* during dry years and remained stable in *P. latifolia* [11]; and fruit production correlated with annual rainfall in *Q. ilex*, but not in *P. latifolia* [11].

Mediterranean vegetation is characterized by the physiological adaptations that allow its tolerance to, and recovery from, episodes of dry environmental conditions; however, if extreme drought events exceed the tipping points of climate tolerance, some sensitive species may fail to survive, whereby species diversity, composition, and associated ecosystem function become permanently altered, leading to perturbations in the ecosystem service provision [13]. *Q. ilex* is the current dominant species of the Mediterranean forest, and is widely distributed across the whole Mediterranean Basin in South Europe and North Africa, and from Portugal to Syria. Under hotter and drier environmental conditions, a decrease in the abundance of *Q. ilex* in Mediterranean forests may be expected, along with a reduction in forest capacity to mitigate the effects of climate change through the sequestration of atmospheric CO_2; there may also be a progressive replacement of *Q. ilex* by other tall shrub species, such as *P. latifolia*, that transforms the structure and ecosystem function of the forest.

In water-limited ecosystems, the principal effects of hydrological processes on vegetation are mediated through soil water content [14]; thus, experimental methods that affect soil moisture effectively simulate the likely impacts of decreases in water availability on plants. Furthermore, long-term climate manipulation experiments in natural ecosystems allow the prediction of the effects of environmental change on ecosystem function and development [15,16] and may highlight the dampening effect on the size of treatments that has been widely reported in global-change experiments [17].

Here, our main objective was to determine the effects of climate change-mediated drought on the development of the Mediterranean forest, paying special attention to the development of the different species and their possible divergent evolution. This study was conducted as a long-term climate manipulation experiment that simulates a 15% decrease in soil moisture, as projected by general circulation models for the Mediterranean region [18].

2. Materials and Methods

2.1. Study Site

The study was conducted in a Mediterranean holm oak forest located on a south facing slope (25%) of the Prades Mountains, Catalonia, in north-eastern Spain (41°20′38″ N, 1°2′0″ E, 950 m a.s.l.), in which the vegetation comprises dense multi-stem, coppiced woody species (16.616 stems ha^{-1}), created for charcoal production; no coppicing activity has taken place in the forest during the preceding 70 years. The dominant species are *Q. ilex* and *P. latifolia* (both species represent more than 80% of the total forest basal area). There are also other evergreen shrub species such as *Arbutus unedo* L., *Erica arborea* L., and *Juniperus oxycedrus* L.; deciduous tree species include the occasional *Sorbus torminalis* L. and *Acer monspessulanum* L. The soils comprise a Dystric Cambisol, and the range in soil depth (35 to 100 cm) results in contrasting levels of water and nutrient availability, reflected by two forest-type structures [19]: a tall canopy forest (8–10 m high) dominated by *Q. ilex*, and a low canopy forest (4–6 m high), with a greater abundance of *P. latifolia* than in tall canopy (Table A1).

2.2. Experimental Design

Eight 15 × 10 m plots were delimited at the same altitude along the slope (four in the tall canopy stand and another four in the low canopy stand) in 1999. Four of these plots (two each in the tall and low canopy stands) were randomly selected as drought plots, and received an experimental partial rainfall exclusion; the four remaining plots were left untreated as a control. In the drought plots, six plastic strips, comprising four rigid PVC tubes connected with transparent PVC, were installed under the canopy along the longitudinal slope, at about 1 m above the soil surface and extending to 2 m outside the plot margin to reduce edge-effects; the strips covered 30% of the plot surface area for the interception of rain throughfall and its removal downslope. Soil moisture was measured monthly using a time domain reflectometer (Delta-T Devices, Cambridge, UK) placed in the upper 15 cm of the soil profile of four randomly selected loci per plot, along with the leaf water potential (measured seasonally, in two leaves from two different trees per species, collected in the upper layer of the canopy and fully exposed to solar radiation) at the same loci using a Scholander-type chamber pressure (PMS Instrument Company, Albany, OR, USA). An automatic meteorological station installed at the study site monitored temperature and rainfall during the 1999–2019 study period.

2.3. Basal Area Determination

In 1999, all of the living tree and shrub stems in the plots were marked at 50 cm above ground level using permanent paint, and their circumference (diameter > 2 cm) was measured at the painted mark annually in winter. The basal area of each stem was determined from the stem circumference measurements, assuming each stem section to be a circle; the annual basal area increment was calculated from the difference in the basal area measured in two consecutive years. For each plot, species, and year, we determined the basal area increment (BAI), basal area lost by stem mortality (BAL), and total basal area increase (TBAI) which was determined as the basal area increment minus basal area lost by stem mortality.

2.4. Data Analysis

Seasonal differences in soil moisture and leaf water potential were tested using repeated measures analysis of variance (ANOVA). January, February and March were combined in "winter", April, May and June were combined in "spring", July, August and September were combined in "summer", and October, November and December were combined in "autumn". The relationships between the BAI, BAL, and TBAI per species and canopy stand (dependent variables) and the annual mean temperature, rainfall, and SPEI climatic indices [3,4,20] (independent variables) were tested using linear and polynomial regression analyses. We used general linear modeling to test the determinant factors of the annual BAI, BAL, and TBAI of *Q. ilex* and *P. latifolia*. The species, treatment and canopy stand were considered independent factors, and the annual mean temperature and rainfall were covariates. Finally, we conducted the same models without the species factor, and including all tree and shrub species present in the plots named "overall species". All analyses were performed using Statistica 10.0 (StatSoft Inc., 2010, Tulsa, OK, USA).

3. Results
3.1. Soil Moisture and Leaf Water Potential

The meteorological conditions of the study site were typically Mediterranean: summers were hot and dry, winters were relatively cold, and rainfall events tended to be concentrated in spring and autumn (Figure 1). Accordingly, both soil moisture and leaf water potential levels showed a clear seasonality; their values decreased in summer, recovered in autumn, and were greatest in spring and winter when evapotranspiration rates were lower. Soil moisture levels were 15% lower in the drought plots than in the control plots ($p < 0.05$) (Figure 2). The decrease in leaf water potential values in summer was stronger in *P. latifolia* than in *Q. ilex* ($p < 0.05$), whereas in autumn, *Q. ilex* maintained a higher water potential than *P. latifolia*; in winter and spring, the levels were similar in both species (Figure 2). The drought treatment did not exert any significant additional effect on the leaf water potential levels.

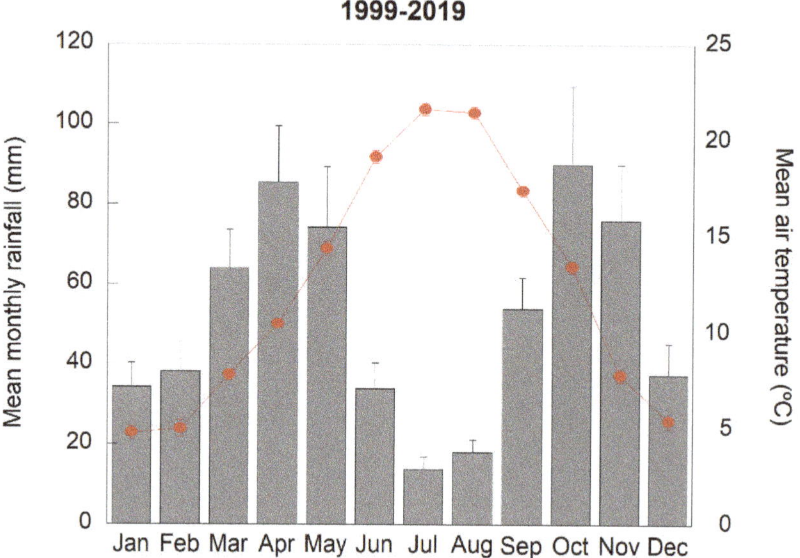

Figure 1. Average meteorological conditions at the study site during the experimental period (1999–2019). Red dots indicate mean monthly air temperature and grey columns indicate mean monthly rainfall. Vertical bars in the grey columns and red dots indicate the standard error of the mean (n = 21 years).

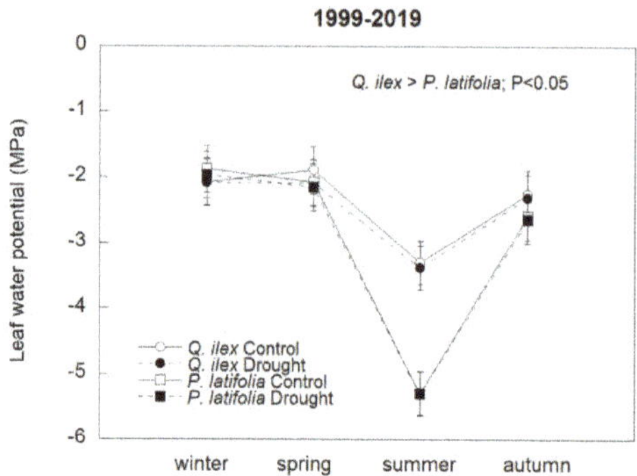

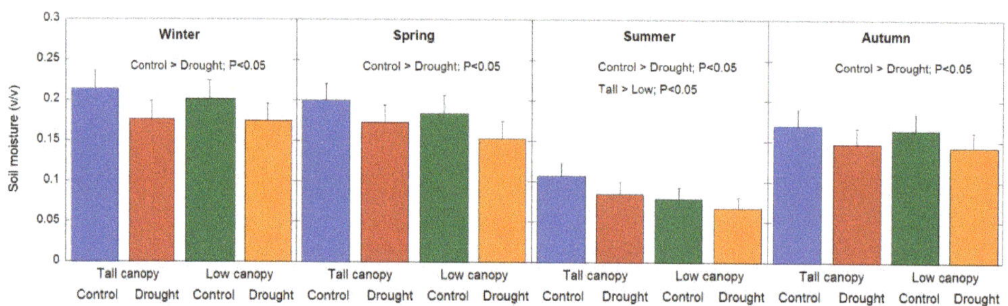

Figure 2. Average seasonal leaf water potential (upper panel) and soil moisture (lower panel) during the study period (1999–2019). Vertical bars indicate the standard error of the mean (n = 21 years).

3.2. Climate Effects on Stem Growth

BAI was positively related to the annual rainfall in the two species separately (*Q. ilex*: $p < 0.05$; *P. latifolia*: $p < 0.01$) and for all of the tree and shrub species ($p < 0.01$) (Figure 3); there was a trend towards a positive relationship between the BAL and the mean annual temperature in *Q. ilex* ($p = 0.06$) and all of the tree and shrub species ($p = 0.06$), but not in *P. latifolia* (Figure 4). As a consequence of lower stem growth in dry years and greater stem mortality in hot years, the TBAI was positively related to the annual rainfall in *Q. ilex* ($p = 0.05$), *P. latifolia* ($p < 0.01$), and all of the tree and shrub species ($p < 0.01$), but these relationships were not linear; they tended to saturate at high precipitation values. On the other hand, there was a negative relationship between the TBAI and the mean annual temperature in *Q. ilex* ($p = 0.05$) and all of the tree and shrub species ($p < 0.05$); in *P. latifolia*, there was a trend towards a negative relationship between the TBAI and the mean annual temperature ($p = 0.06$). Also, these relationships were not linear; they tended to increase at very low temperature values, but strongly decrease at high temperatures. Thes effects of climate on stem growth differed between the two canopy stands, where the mean annual temperature tended to be more determinant for the TBAI in the tall canopy stand for *Q. ilex* ($p < 0.05$) and all of the tree and shrub species together ($p < 0.05$), whereas the annual rainfall was more determinant for the TBAI in the low canopy stand for *Q. ilex* ($p = 0.05$), *P. latifolia* ($p < 0.01$) and all of the tree and shrub species together ($p < 0.05$) (Figure 5). However, we did not obtain significative results between the BAI measurements and SPEI climatic indices.

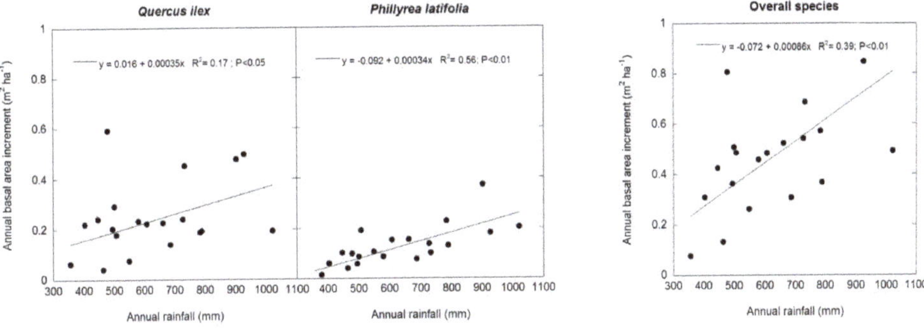

Figure 3. Linear regression analysis of annual basal area increment (BAI) and annual rainfall in *Q. ilex*, *P. latifolia*, and all tree and shrub species. $n = 21$ years.

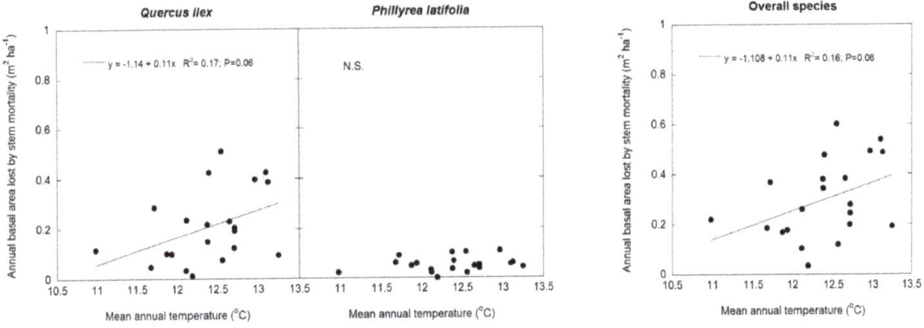

Figure 4. Linear regression analysis of annual basal area lost by stem mortality (BAL) and mean annual temperature in *Q. ilex*, *P. latifolia*, and all tree and shrub species. $n = 21$ years.

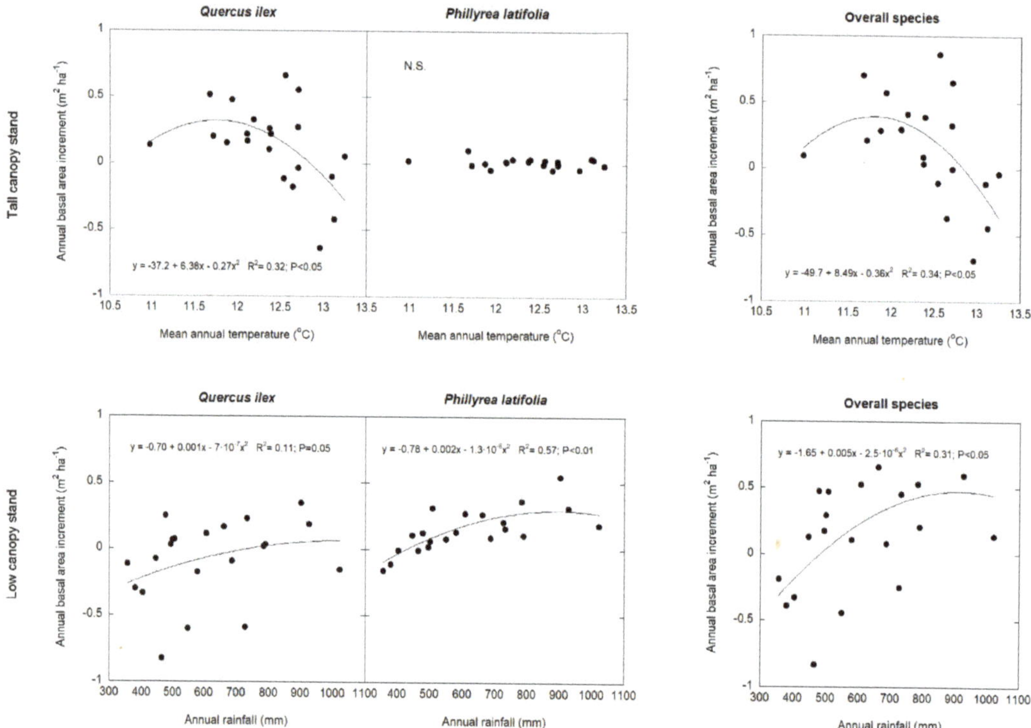

Figure 5. Polynomial regression analysis of total annual basal area increase (TBAI) and mean annual temperature in the tall canopy stand and total annual rainfall in the low canopy stand, for *Q. ilex*, *P. latifolia*, and all of the tree and shrub species. *n* = 21 years.

3.3. Drought Effects on Stem Growth

Between 1999 and 2019, regardless of treatment, the BAI in *Q. ilex* was greater in the tall canopy stand than in the low canopy stand, while in *P. latifolia* it was lower in the tall canopy stand than in the low canopy stand (species × canopy interaction: $F = 114.1$, $p < 0.01$), reflecting the recorded basal area of the species in 1999 and 2019 (Table A1); there was no difference in the BAI between the tall and low canopy stands when all of the tree and shrub species were considered together (Figure 6). While there were no effects of the drought on the BAI in *Q. ilex*, *P. latifolia*, or all of the tree and shrub species, the BAL was greater in the drought plots in *Q. ilex* and *P. latifolia* ($F = 8.4$, $p < 0.01$), and all of the tree and shrub species ($F = 8.1$, $p < 0.01$), with no overall BAL canopy type differences in *Q. ilex* or all of the tree and shrub species. In *P. latifolia*, the BAL was greater in the low canopy stand (Figure 6); however, the BAL was always higher in *Q. ilex* than in *P. latifolia* ($F = 33.9$, $p < 0.01$). Across the study period, the TBAI reflected clear differences between the two species: *Q. ilex* experienced a higher TBAI in the tall canopy stand, whereas *P. latifolia* experienced higher TBAI in low canopy stand (species x canopy interaction: $F = 29.1$, $p < 0.01$); there was also a clear decrease in the TBAI levels induced by drought ($F = 5.7$, $p = 0.02$), especially in *Q. ilex* (species x treatment interaction: $F = 5.1$, $p = 0.02$). We found that the maximum TBAI in *Q. ilex* was recorded in non-restricted rainfall (control) plots located in the tall canopy stand, where there was a slight positive association between the TBAI and drought treatment in the tall canopy stand and a slight negative association between the TBAI and non-restricted rainfall in the low canopy stand; the lowest TBAI was recorded in drought plots of the low canopy stand (Figure 7). When all of the tree and shrub species were considered together, there was a decrease in the TBAI levels in drought plots ($F = 12.9$, $p < 0.01$) and there was no difference in the canopy stand type (Figure 7).

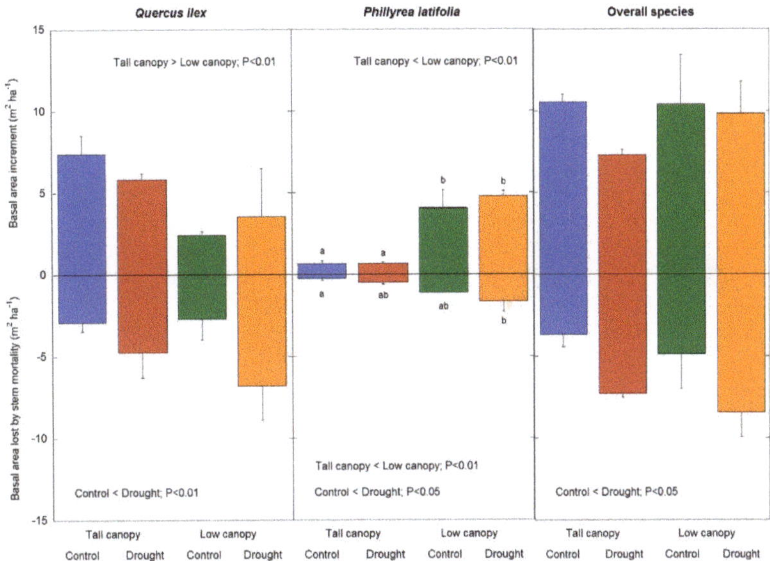

Figure 6. Drought treatment and canopy stand type effects on basal area increment (BAI) and basal area lost by stem mortality (BAL) in *Q. ilex*, *P. latifolia*, and all of the tree and shrub species during the study period (1999–2019). Vertical bars indicate the standard error of the mean ($n = 21$ years). Different letters indicate drought treatment differences between and within the canopy stand types ($p < 0.05$).

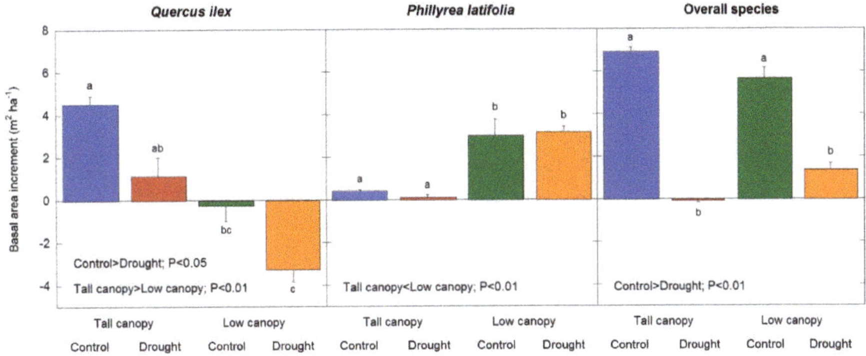

Figure 7. Drought treatment and canopy stand type effects on total basal area increment (TBAI) in *Q. ilex*, *P. latifolia*, and all of the tree and shrub species during the study period (1999–2019). Vertical bars indicate the standard error of the mean ($n = 21$ years). Different letters indicate drought treatment differences between and within the canopy stand types ($p < 0.05$).

Overall, the TBAI of the most representative species of the study forest followed three main response patterns to changes in soil moisture: species with a greater TBAI under non-drought conditions, such as *Q. ilex*, A. unedo, and J. oxycedrus; species with a similar TBAI, regardless of drought conditions, such as *P. latifolia*, E. arborea, and Rhamnus alaternus; and species with decreases in the TBAI, regardless of drought conditions, albeit with larger decreases under drought conditions, such as S. torminalis (Figure 8).

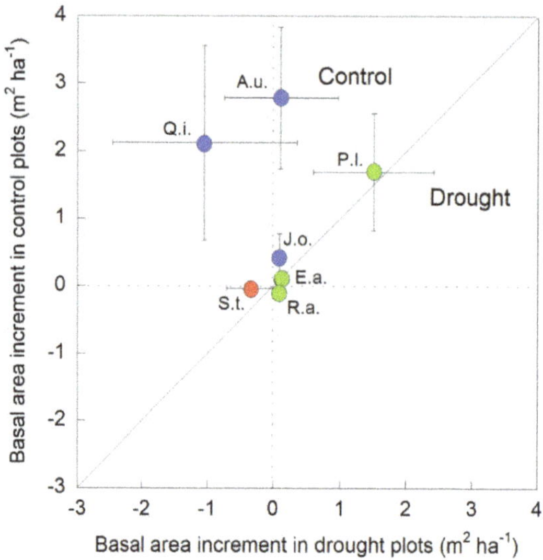

Figure 8. Total basal area increment (TBAI) of the most representative forest tree and shrub species under full (control) and partial (drought) rainfall conditions during the study period (1999–2019). Vertical and horizontal bars indicate the standard error of the mean (n = 21 years). Species with a higher TBAI in control plots than in drought plots are depicted with blue circles, species with a similar TBAI in both control and drought plots are depicted with green circles, and the species with a negative TBAI in both the control and drought plots are depicted with a red circle. Q.i. means *Quercus ilex*, A.u. means *Arbutus unedo*, J.o. means *Juniperus oxycedrus*, P.l. means *Phillyrea latifolia*, E.a. means *Erica arborea*, R.a. means *Rhamnus alaternus*, and S.t. means *Sorbus torminalis*.

4. Discussion

As observed previously at this experimental study site [3,4,12,20], we found that the tree and shrub stem growth was constrained by water availability and tree mortality increased in dry and hot years. These effects were particularly marked in the dominant species, *Q. ilex*, whereas in other species, such as *P. latifolia*, that are more tolerant of lower levels of water availability, there were no effects of experimental drought and there was positive stem growth, especially where the *Q. ilex* abundance was declining in the droughted low canopy stand. Many studies report an increase in basal area increment with tree size [21], and a different climate sensitivity depending on tree size [22], but the *Q. ilex* stem size tended to be greater than that of *P. latifolia*. Despite this, the BAI was lower in *Q. ilex* than in *P. latifolia*, and in the low canopy stand where the tree size of both of the species was similar, climate sensitivity was higher in *Q. ilex* than in *P. latifolia*. These two species share a strategy to avoid excessive water loss due to transpiration; they close their stomata when they are subjected to a high vapor pressure deficit and low water availability. However, previous findings from this long-term experiment revealed only a small decrease in net photosynthetic CO_2 uptake induced by the drought treatment in the two species; this small reduction in the photosynthetic rate is insufficient to explain the considerable decrease in the BAI observed in *Q. ilex* from the drought plots [23]. Changes in the leaf water potential and other physiological parameters may explain the lower BAI in droughted *Q. ilex* in this study; for example, lower production of new leaves in *Q. ilex* canopies has been recorded under drought conditions than under control conditions, with no such treatment differences in the *P. latifolia* leaf demography [24].

We found that the experimental drought increased stem mortality in all of the tree and shrub species, but this effect was stronger in *Q. ilex* than in *P. latifolia*. Increases in both the air temperature and dry conditions are considered to be the most important drivers of tree mortality due to hydraulic failure and carbon starvation [6,7,25]. In our study, mortality increased markedly with the mean annual temperature in *Q. ilex*, whereas *P. latifolia* was more tolerant of higher temperatures, as indicated by the similar mortality rates across the study years [12]. Despite their similar physiological strategies to increase tolerance of dry conditions, detailed xylem architecture observations revealed wider xylem vessels in *Q. ilex* than in *P. latifolia* [26] that allow the transport of greater amounts of sap under well-watered conditions, but are more susceptible to embolism under drought conditions.

Shifts in species distributions induced by climate change are expected to occur globally in several biomes [27,28]. Across the duration of this long-term experiment, the total basal area of *Q. ilex* decreased exponentially with an increase in air temperature and decrease in rainfall, while a previous study showed the contrasting presence of abundant *P. latifolia* seedlings and young individuals and scarce *Q. ilex* seedlings in this forest [29], indicating that the replacement of moribund *Q. ilex* may only be possible from resprouting shoots. These results indicate that the warming air temperatures and decreasing water availability that are predicted for the Mediterranean Basin [1] will lead to a receding *Q. ilex* distribution and progressive substitution by shrub species better adapted to drier conditions, particularly under extreme arid conditions, such as those in the low canopy stand. By the first time, we detected in our experimental site a clear substitution of *Q. ilex* by other shrub species, as it was clearly observed in the low canopy stand.

Across the study period, the total basal area of the drought plots did not increase, and this reduction in plant productivity was accompanied by other important changes in the ecosystem function of the forest. For example: leaf material from the drought treatment contained greater concentrations of nitrogen (N)-15 than that from the control, due to greater ecosystem N losses under the experimental drought conditions [30]; bacteria diversity has been shown to be particularly sensitive to water availability and fungi diversity was more tolerant to seasonal and experimental changes in soil moisture [31]; and drought conditions led to decreases in soil phosphatase activity [32] and altered trophic interactions, such as increases in leaf sugar and phenol concentrations to resist drier conditions, that boosted levels of insect herbivory [19].

In order to increase ecosystem contributions to climate change mitigation, considerable efforts have been made to preserve species composition and the associated ecosystem function and service provision [33]. For Mediterranean forests, such as that studied here, with high levels of stem density and competition and an associated low BAI [34], the reduction of stem density by selective thinning may represent an appropriate management option to improve the BAI [12,35].

5. Conclusions

The results from this long-term study demonstrated a progressive replacement of the dominant species, *Q. ilex*, by another shrub species, *P. latifolia*, that was better adapted to the hotter and drier climatic conditions that are predicted to impact this Mediterranean region over the coming decades. The slight decrease in water availability, as simulated by the drought treatment, that reduced soil moisture by 15% exacerbated this transition in relative species abundance and contributed to a decrease in stem growth and an increase in tree mortality in *Q. ilex*. The experimental drought also elicited additional, ecosystem level impacts, including losses in N, changes in the metabolic activity of soil organisms, and the capacity to resist attack by plant pathogens. Importantly, some ecosystem services of this forest type, such as the amelioration of climate change impacts as a carbon sink, are expected to decline further with the ongoing and continued changes in the climate.

Author Contributions: Conceptualization, experimental design, statistical analyses, and discussion of the results: R.O. and J.P.; writing: R.O.; and, supervision: J.P. All authors have read and agreed to the published version of the manuscript.

Funding: This research was financially supported by the Spanish government project PID2019-110521GB-I00 ELEMENTAL SHIFT, the European Research Council Synergy grant ERC-2013-SyG-2013-610028 IMBALANCE-P, and the Catalan government grant SGR-2017-1005.

Acknowledgments: This study was undertaken with the approval of Ester Trullols, director of the PNIN de Poblet Natural Park.

Conflicts of Interest: The authors declare no conflict of interest.

Appendix A

Table A1. Mean ($\pm$SEM) basal area (m^2 ha^{-1}) of *Q. ilex*, *P. latifolia*, and all tree and shrub species in the plot types at the start and the end of the study period (1999–2919) (n = 2 plots).

Species	Plot Type	Basal Area 1999	Basal Area 2019
Overall Species	Tall control	45.83 $\pm$ 2.89	52.74 $\pm$ 2.58
Overall Species	Tall drought	48.68 $\pm$ 3.87	48.64 $\pm$ 4.02
Overall Species	Low control	39.19 $\pm$ 9.37	44.79 $\pm$ 10.18
Overall Species	Low drought	33.25 $\pm$ 9.39	34.59 $\pm$ 9.86
Quercus ilex	Tall control	37.70 $\pm$ 3.67	42.18 $\pm$ 4.24
Quercus ilex	Tall drought	37.94 $\pm$ 4.91	39.08 $\pm$ 3.68
Quercus ilex	Low control	15.99 $\pm$ 4.92	15.75 $\pm$ 3.87
Quercus ilex	Low drought	19.03 $\pm$ 12.14	15.78 $\pm$ 12.97
Phillyrea latifolia	Tall control	2.19 $\pm$ 1.14	2.61 $\pm$ 1.23
Phillyrea latifolia	Tall drought	2.88 $\pm$ 0.48	2.94 $\pm$ 0.23
Phillyrea latifolia	Low control	15.24 $\pm$ 3.37	18.20 $\pm$ 4.49
Phillyrea latifolia	Low drought	13.21 $\pm$ 2.65	16.36 $\pm$ 2.29

References

1. Allen, M.R.; Babiker, M.; Chen, Y.; de Coninck, H.; Connors, S.; van Diemen, R.; Dube, O.P.; Ebi, K.L.; Engelbrecht, F.; Ferrat, M.; et al. IPCC 2018: Summary of Policymakers. In *Global Warming of 1.5 °C. An IPCC Special Report on the Impacts of Global Warming of 1. 5 °C Above Pre-Industrial Levels and Related Global Greenhouse Gas Emission Pathways, in the Context of Strengthening the Global Response to the Threat of Climate Change, Sustainable Development, and Efforts to Eradicate Poverty*; Masson-Delmotte, V., Pörtner, H.O., Skea, J., Zhai, P., Roberts, D., Shukla, P.R., Pirani, A., Moufouma-Okia, W., Péan, C., Pidcock, R., et al., Eds.; World Meteorological Organization: Geneva, Switzerland, 2018; p. 32.
2. Wang, W.; Peng, C.; Kneeshaw, D.D.; Larocque, G.R.; Luo, Z. Drought induced tree mortality: Ecological consequences, causes, and modeling. *Environ. Rev.* **2012**, *20*, 109–121. [CrossRef]
3. Barbeta, A.; Ogaya, R.; Peñuelas, J. Dampening effects of long-term experimental drought on growth and mortality rates of a Holm oak forest. *Glob. Chang. Biol.* **2013**, *19*, 3133–3144. [CrossRef]
4. Liu, D.; Ogaya, R.; Barbeta, A.; Yang, X.; Peñuelas, J. Long-term experimental drought combined with natural extremes accelerate vegetation shift in a Mediterranean holm oak forest. *Environ. Exp. Bot.* **2018**, *151*, 1–11. [CrossRef]
5. Carnicer, J.; Coll, M.; Ninyerola, M.; Pons, X.; Sanchez, G.; Peñuelas, J. Widespread crown condition decline, food web disruption, and amplified tree mortality with increased climate change-type drought. *Proc. Natl. Acad. Sci. USA* **2011**, *108*, 1474–1478. [CrossRef]
6. Allen, C.D.; Macalady, A.K.; Chenchouni, H.; Bachelet, D.; McDowell, N.; Vennetier, M.; Kitzberger, T.; Rigling, A.; Breshears, D.D.; Hogg, E.H.; et al. A global overview of drought and heat-induced tree mortality reveals emerging climate change risks for forests. *For. Ecol. Manag.* **2010**, *259*, 660–684. [CrossRef]
7. Breshears, D.D.; Cobb, N.S.; Rich, P.M.; Price, K.D.; Allen, C.D.; Balice, R.G.; Romme, W.H.; Kastens, J.H.; Floyd, M.L.; Belnap, J.; et al. Regional vegetation die-off in response to global-change-type drought. *Proc. Natl. Acad. Sci. USA* **2005**, *102*, 15144–15148. [CrossRef] [PubMed]
8. Moore, D.J.P.; Aref, S.; Ho, R.M.; Pippen, J.S.; Hamilton, J.G.; de Lucia, E.H. Annual basal area increment and growth duration of *Pinus taeda* in response to eight years of free-air carbon dioxide enrichment. *Glob. Chang. Biol.* **2006**, *12*, 1367–1377. [CrossRef]
9. Terradas, J. Holm oak and holm oak forests: An introduction. In *Ecology of Mediterranean Evergreen oak Forests*; Rodà, F., Retana, J., Gracia, C.A., Bellot, J., Eds.; Springer: Berlin, Germany, 1999; pp. 3–14.
10. Peñuelas, J.; Filella, I.; Llusià, J.; Siscart, D.; Piñol, J. Comparative field study of spring and summer leaf gas Exchange and photobiology of the Mediterranean trees *Quercus ilex* and *Phillyrea latifolia*. *J. Exp. Bot.* **1998**, *49*, 229–238. [CrossRef]
11. Ogaya, R.; Penuelas, J. Wood vs. canopy allocation of aboveground net primary productivity in a Mediterranean forest during 21 years of experimental rainfall exclusion. *Forests* **2020**, *11*, 1094. [CrossRef]

12. Ogaya, R.; Liu, D.; Barbeta, A.; Peñuelas, J. Stem Mortality and Forest Dieback in a 20-Years Experimental Drought in a Mediterranean Holm Oak Forest. *Front. For. Glob. Chang.* **2020**, *2*, 89. [CrossRef]

13. Peñuelas, J.; Sardans, J.; Filella, I.; Estiarte, M.; Llusià, J.; Ogaya, R.; Carnicer, J.; Bartrons, M.; Rivas-Ubach, A.; Grau, O.; et al. Impacts of global change on Mediterranean forests and their services. *Forests* **2017**, *8*, 463. [CrossRef]

14. Yin, J.; D'Odorico, P.; Porporato, A. Soil Moisture Dynamics in Water-Limited Ecosystems. In *Dryland Ecohydrology*; D'Odorico, P., Porporato, A., Wilkinson Runyan, C., Eds.; Springer: Cham, Switzerland, 2019; pp. 31–48.

15. Kröel-Dulay, G.; Ransijn, J.; Schmidt, I.K.; Beier, C.; de Angelis, P.; de Dato, G.; Dukes, J.S.; Emmett, B.A.; Estiarte, M.; Garadnai, J.; et al. Increased sensitivity to climate change in disturbed ecosystems. *Nat. Comm.* **2015**, 1–7. [CrossRef]

16. Liu, D.; Zhang, C.; Ogaya, R.; Estiarte, M.; Peñuelas, J. Effects of decadal experimental drought and climate extremes on vegetation growth in Mediterranean forests and shrublands. *J. Veg. Sci.* **2020**, *31*, 768–779. [CrossRef]

17. Leuzinger, S.; Luo, Y.; Beier, C.; Dieleman, W.; Vicca, S.; Körner, C. Do global change experiments overestimate impacts on terrestrial ecosystems? *Trends Ecol. Evol.* **2011**, *26*, 236–241. [CrossRef]

18. Bates, B.C.; Kundzewicz, Z.W.; Wu, S.; Palutikof, J.P. *Climate Change and Water. Technical Paper of the Intergovernmental Panel on Climate Change*; IPCC Secretariat: Geneva, Switzerland, 2018; p. 210.

19. Rivas-Ubach, A.; Gargallo-Garriga, A.; Sardans, J.; Oravec, M.; Mateu-Castell, L.; Pérez-Trujillo, M.; Parella, T.; Ogaya, R.; Urban, O.; Peñuelas, J. Drought enhances folivory by shifting foliar metabolomes in *Quercus ilex* trees. *New Phytol.* **2014**, *202*, 874–885. [CrossRef]

20. Ogaya, R.; Barbeta, A.; Basnou, C.; Peñuelas, J. Satellite data as indicators of tree biomass growth and forest dieback in a Mediterranean holm oak forest. *Ann. For. Sci.* **2015**, *72*, 135–144. [CrossRef]

21. Stephenson, N.L.; Das, A.J.; Condit, R.; Russo, S.E.; Baker, P.J.; Beckman, N.G.; Coomes, D.A.; Lines, E.R.; Morris, W.K.; Rüger, N.; et al. Rate of tree carbon accumulation increases continuously with tree size. *Nature* **2014**, *507*, 90–93. [CrossRef]

22. Trouillier, M.; van der Maaten-Theunissen, M.; Scharnweber, T.; Würth, D.; Burger, A.; Schnittler, M.; Wilmking, M. Size matters–a comparison of three methods to assess age- and size-dependent climate sensitivity of trees. *Trees* **2019**, *33*, 183–192. [CrossRef]

23. Ogaya, R.; Llusià, J.; Barbeta, A.; Asensio, D.; Liu, D.; Alessio, G.A.; Peñuelas, J. Foliar CO_2 in a holm oak forest subjected to 15 years of climate change simulation. *Plant Sci.* **2014**, *226*, 101–107. [CrossRef] [PubMed]

24. Ogaya, R.; Peñuelas, J. Contrasting foliar responses to drought in *Quercus ilex* and *Phillyrea latifolia*. *Biol. Plant.* **2006**, *50*, 373–382. [CrossRef]

25. Sevanto, S.; McDowell, N.G.; Dickman, L.T.; Pangle, R.; Pockman, W.T. How do trees die? A test of the hydraulic failure and carbon starvation hypotheses. *Plant Cell Environ.* **2014**, *37*, 153–161. [CrossRef] [PubMed]

26. Martínez-Vilalta, J.; Prat, E.; Oliveras, I.; Piñol, J. Hydraulic properties of roots and stems of nine woody species from a Holm oak forest in NE Spain. *Oecologia* **2002**, *133*, 19–29. [CrossRef] [PubMed]

27. Wu, Z.; Dijskstra, P.; Koch, G.W.; Peñuelas, J.; Hungate, B.A. Responses of terrestrial ecosystems to temperature and precipitation change: A meta–analysis of experimental manipulation. *Glob. Chang. Biol.* **2011**, *17*, 927–942. [CrossRef]

28. Peñuelas, J.; Sardans, J.; Filella, I.; Estiarte, M.; Llusià, J.; Ogaya, R.; Carnicer, J.; Bartrons, M.; Rivas-Ubach, A.; Grau, O.; et al. Assessment of the impacts of climate change on Mediterranean terrestrial ecosystems based on data from field experiments and long-term monitored field gradients in Catalonia. *Env. Exp. Bot.* **2018**, *152*, 49–59.

29. Lloret, F.; Peñuelas, J.; Ogaya, R. Seedling versus sprout recruitment of two Mediterranean trees (*Phillyrea latifolia* and *Quercus ilex*) under drought conditions. *J. Veg. Sci.* **2004**, *15*, 237–244. [CrossRef]

30. Ogaya, R.; Peñuelas, J. Changes in leaf $\delta^{13}C$ and $\delta^{15}N$ and trunk growth for three Mediterranean tree species in relation to soil water availability. *Acta Oecol.* **2008**, *34*, 331–338. [CrossRef]

31. Curiel-Yuste, J.; Peñuelas, J.; Estiarte, M.; Garcia-Mas, J.; Mattana, S.; Ogaya, R.; Pujol, M.; Sardans, J. Drought-resistant fungi control soil organic matter decomposition and its response to temperature. *Glob. Chang. Biol.* **2011**, *17*, 1475–1486. [CrossRef]

32. Sardans, J.; Peñuelas, J.; Ogaya, R. Experimental drought reduced acid and alkaline phosphatise activity and increased organic extractable P in soil in a *Quercus ilex* Mediterranean forest. *Eur. J. Soil Biol.* **2008**, *44*, 509–520. [CrossRef]

33. Ingrisch, J.; Bahn, M. Towards a comparable quantification of resilience. *Trends Ecol. Evol.* **2018**, *33*, 251–259. [CrossRef] [PubMed]

34. Vospernik, S. Basal area increment models accounting for climate and mixture for Austrian tree species. *For. Ecol. Manag.* **2021**, *480*, 118725. [CrossRef]

35. Martín-Benito, D.; Rio, M.; Heinrich, I.; Helle, G.; Cañellas, I. Response of climate-growth relationships and water use efficiency to thinning in a *Pinus nigra* afforestation. *For. Ecol. Manag.* **2010**, *259*, 967–975. [CrossRef]

forests

MDPI

Article

Sap Flow in Aleppo Pine in Greece in Relation to Sapwood Radial Gradient, Temporal and Climatic Variability

Evangelia Korakaki [1] and Mariangela N. Fotelli [2,*]

[1] Institute of Mediterranean and Forest Ecosystems, Hellenic Agricultural Organization Demeter, P. O. Box 14180, Terma Alkmanos, Ilisia, 11528 Athens, Greece; e.korakaki@fria.gr

[2] Forest Research Institute, Hellenic Agricultural Organization Demeter, Vassilika, 57006 Thessaloniki, Greece

* Correspondence: fotelli@fri.gr; Tel.: +30-2310-461172 (ext. 238)

Abstract: Research Highlights: The radial gradient of sap flux density (Js) and the effects of climatic factors on sap flow of Aleppo pine were assessed at different time scales in an eastern Mediterranean ecosystem to improve our understanding of the species water balance. Background and Objectives: Aleppo pine's sap flow radial profile and responses to environmental parameters in the eastern Mediterranean were, to our best knowledge, originating to date from more arid planted forests. Information from natural forests in this region was lacking. Our objectives were to (a) determine the species' radial variability in Js on a diurnal and seasonal basis and under different climatic conditions, (b) scale up to tree sap flow taking into account the radial profile of Js and (c) determine the responses of Aleppo pine's sap flow over the year to climatic variability. Materials and Methods: Js was monitored in Aleppo pine in a natural forest in northern Greece with Granier's method using sensors at three sapwood depths (21, 51, and 81 mm) during two periods differing in climatic conditions, particularly in soil water availability. Results: Js was the highest at 21 mm sapwood depth, and it declined with increasing depth. A steeper gradient of Js in deep sapwood was observed under drier conditions. The same patterns of radial variability in Js were maintained throughout the year, but the contribution of inner sapwood to sap flow was the highest in autumn when the lower seasonal Js was recorded in both study periods. Not taking into account the radial gradient of Js in the studied Aleppo pine would result in a c. 20.2–27.7 % overestimation of total sap flow on a sapwood basis (Qs), irrespective of climatic conditions. On a diurnal and seasonal basis, VPD was the strongest determinant of sap flux density, while at a larger temporal scale, the effect of soil water content was evident. At SWC > 20% sap flow responded positively to increasing solar radiation and VPD, indicating the decisive role of water availability in the studied region. Moreover, in drier days with VPD > 0.7 KPa, SWC controlled the variation of sap flow. Conclusions: There is a considerable radial variability in Js of the studied Aleppo pine and a considerable fluctuation of sap flow with environmental dynamics that should be taken into account when addressing the species water balance.

Keywords: sap flux; radial profile; sapwood depth; Aleppo pine; diurnal variation; seasonal variation; climate

Citation: Korakaki, E.; Fotelli, M.N. Sap Flow in Aleppo Pine in Greece in Relation to Sapwood Radial Gradient, Temporal and Climatic Variability. *Forests* **2021**, *12*, 2. https://dx.doi.org/10.3390/f12010002

Received: 23 November 2020
Accepted: 18 December 2020
Published: 22 December 2020

Publisher's Note: MDPI stays neutral with regard to jurisdictional claims in published maps and institutional affiliations.

1. Introduction

Sap flux density and sap flow measurements are widely used in ecophysiological studies to investigate aspects of plant–water interactions such as canopy conductance [1–5], whole-plant hydraulic conductance [3,6], and leaf stomatal conductance [3,6–9]. To estimate these variables from sap flow monitoring, sap flux density (Js) measurements need to be upscaled to the tree level. In several studies [10–17], sap flow is estimated from sap flux density measurements with sensors installed in the outer part of the xylem, assuming uniform sap flux density across the sapwood. Berdanier et al. [18] reported that out of 122 reviewed articles (published between 2013 and 2015) that scaled

51

up sap flux observations to whole-tree or stand levels, 58% assumed homogenous flow throughout the xylem.

However, neglecting radial variability may lead to systematic errors in the total sap flow [19–25]. Čermak and Nadezhdina [20] showed that upscaling sap flux density, measured in the outer half of the sapwood of different species, led to overestimation of total tree sap flow (by 10–140%), whereas when measured in deep inner sapwood layers, total tree sap flow was underestimated (by 40–80%). Similarly, Delzon et al. [22] showed that ignoring the variability in radial profile of sap flow in maritime pine trees led to an overestimation of transpiration and stand water use that increased with increasing stem diameter from 5% for the young stands to 47% for the oldest stands. To overcome these uncertainties, the radial variability in sap flux density in different trees has been determined [19,26–30] or correction factors accounting for the spatial variation of sap flow with increasing sapwood radius have been developed [23,31]. Therefore, taking into consideration radial variability is highly important when upscaling sap flux density measurements to accurately estimate tree and stand level water balance, particularly for tracheid-bearing conifers that tend to have deep functional sapwood [18,32].

Variation in the sap flux radial profile in conifer stems, as well as variation in sap flow on the tree level, may be associated with time-dependent changes in environmental conditions. A number of studies have reported that changes in radial patterns were related to environmental variables, such as vapor pressure deficit and soil water availability. Phillips et al. [33] indicated that nighttime recharge of stem water storage increased with increasing soil moisture depletion. Čermak and Nadezhdina [20] showed that, under long-term drought, deeper layers of the sapwood contributed more to water transport. Similarly, Ford et al. [27] found that the radial profile of sap flux density was related to daily changes in evaporative demand, and that during periods of high transpiration, the inner xylem contributed more to total tree sap flow. Hernandez-Santana et al. [34] observed that Js increased at all explored xylem depths with increasing VPD from 8:00 to 11:00 GMT, but with higher VPD (till 15:00 GMT), Js continued increasing only in the deepest xylem, whereas the outer one decreased. In the study of Sanchez-Costa et al. [31], *P. halepensis* showed higher sensitivity of Js to VPD, than the rest diffuse-porous species examined and a higher soil moisture threshold for Js reductions, consistent with the species strict stomatal control under drought. Finally, Cohen et al. [35] supported that the depth of conducting sapwood in *P. halepensis* trees was larger in trees growing under a semi-arid climate than under a sub-humid climate. Thus, assessment of the climatic controls in different time scales on sap flow patterns is important in our effort to understand and predict Aleppo pine's responses to climate variability.

Aleppo pine is widely distributed in coastal and low elevation areas of the Mediterranean basin [36,37], being predominately threatened by the prevailing climate change [38], particularly at its eastern limits [39]. Thus, assessing this species' water balance in relation to environmental factors is crucial. Still, to our knowledge, limited information is available about the radial patterns of sap flux and its seasonal and diurnal fluctuation, in relation to climate, in *P. halepensis* at eastern Mediterranean ecosystems. Mostly, if not solely, the relevant literature refers to Aleppo pine plantations in Israel, such as the works of Schiller and Cohen [40] who first reported on the radial profile of sap flux in this species and Cohen et al. [35] who studied the radial sap flux gradients of Aleppo pines under different climatic conditions.

In the present study, we assessed the diurnal and seasonal variability of the radial profile of sap flux density of a mature near-coastal Aleppo pine ecosystem in Chalkidiki, northern Greece. We measured sap flux density in three sapwood depths (21, 51 and 81 mm) to take into account the radial variability of sap flux of Aleppo pine trees in estimating total sap flow and to explore its response to different temporal scales and to climatic variability. Given that high temporal variations in the sap flux density radial profile may be observed (e.g., [26]), monitoring was conducted during two periods (July 2008–November 2009 and April 2019–April 2020) differing in climatic conditions and predominantly in water availability.

Our specific aims were to: (a) determine the radial profile of Js in Aleppo pine and explore its changes at a diurnal and seasonal scale during two periods differing in climatic conditions, (b) scale up to tree sap flow based on the radial variability of Js and quantify errors that occur when single-depth Js is scaled up to tree level and (c) assess the responses of Aleppo pine sap flow over the year to climatic variability.

2. Materials and Methods

2.1. Description of Study Site and Data Collection

The study was conducted in a natural Aleppo pine (*Pinus halepensis* Mill.) stand located at the Stavronikita forest, Chalkidiki, Northern Greece (latitude: 40°06′ N, longitude: 23°18′ E). The Aleppo pine stand grows at an elevation of 15 m above the sea level, has a mean slope 1% and is located at about 300 m distance from the coast. The understorey consists of a maquis shrub vegetation, dominated with *Pistacia lentiscus* L., *Phyllirea media* L. and *Quercus coccifera* L. The soil has a high pH (7.5–8.2) and lies at the boundary between Calcari-chromic Vertisols and Chromic Luvisols. For the period 2008–2019, mean, minimum and maximum air temperatures were 16.3 °C, 30.3 °C, and 3.5 °C, respectively, while mean annual rainfall was 604.8 mm. A xerothermic season occurred from June till September.

Data were collected during two periods differing in soil water content and other climatic conditions, in order to determine variations in sap flow parameters (sap flux radial gradient-Js and sap flow per sapwood area -Qs) under different climatic conditions. The first recording period was from 1 July 2008 to 30 November 2009 and it was characterized by lower soil water content (hereafter called "dry period"), while the second period lasted from 13 April 2018 to 12 April 2020 and was characterized by higher soil water content (hereafter called "wet period").

2.2. Micrometeorological Parameters

Micrometeorological data were continuously recorded at a fully automated weather station operating at a c. 50 m distance from the forest stand. The recorded parameters included air temperature and air relative humidity at 5 m height (RHT2nl, Delta-T Devices Ltd., Cambridge, UK), solar radiation (SKS1110, Skye Instruments, Llandrindod Wells, UK), wind speed (model 4.3515.30.000, THIES CLIMA, Göttingen, Germany), wind direction (WD4, Delta-T Devices Ltd., Cambridge, UK), precipitation (AR100 and RGB1, EML, North Shields, UK), and soil water content at a 30 cm depth (ML2X/W, Delta-T Devices Ltd., Cambridge, UK). All values were data-logged (DL2e Delta-T Logger, Delta-T Devices Ltd., Cambridge, UK) at 60-min and 30-min intervals during the 1st and the 2nd period, respectively. Vapor pressure deficit (VPD) was also calculated on a 60-min and 30-min basis from relative humidity and air temperature data.

2.3. Biometric Traits of Selected Trees and Sapwood Area Determination

The mean Aleppo pine tree density at the study site is c. 130 trees/ha. The majority of the trees (c. 85%) have a DBH > 30 cm. In order to measure the sap flux of trees that are representative of the studied forest, individuals with DBH lower than this threshold were excluded, while attention was paid to selecting non-neighboring, dominant Aleppo pines. The biometric traits of the selected Aleppo pines are given in Table 1. DBH and sapwood depth were measured during both study periods, while tree height and canopy projected area were determined only during the 2nd (wet) period. For the determination of canopy projected area, the radius of the crown's orthogonal projection was measured at four directions (north, south, east, and west) to build an average radius per tree. One Aleppo pine tree (Pine 4) has been harvested after the 1st (dry) period and has been replaced by another one with similar DBH and height, before the initiation of the 2nd period, as indicated in Table 1.

Table 1. Biometric traits of the monitored Aleppo pine trees.

Pine	DBH (cm)		Sapwood Radius (cm)		Height (m)	Canopy Projected Area (m^2)
	1st period	2nd period	1st period	2nd period		
1	56.3	57.1	25.0	25.3	19.4	100.3
2	43.9	45.8	19.7	20.5	17.4	48.4
3	33.7	37.1	15.3	16.8	14.9	40.2
4 [1]	30.6	34.1	14.0	15.5	18.4	36.3
5	51.9	53.8	23.1	23.9	21.3	98.5

[1] Pine 4 was harvested after the 1st period and replaced by another one with similar DBH and height, before the 2nd period.

Sapwood depth was estimated with an allometric relationship derived from wood slices from 20 Aleppo pines from a close-by Aleppo pine forest (in Pefkochori, Chalkidiki; 39°58′26″ N, 23°36′34″ E) as described in Fotelli et al. [41]. The equation best describing the relationship between sapwood area and diameter at breast height (DBH) ($R^2 = 0.999$, $p < 0.05$) was:

$$As = 0.077 \times DBH^{1.9905}$$

where As stands for sapwood area (in m^2) and DBH for diameter at breast height (in m). To verify this allometric equation for the studied forest stand, DBH and As were measured in additional eight (8) Aleppo pine trees from the study site, and the relationship between them was tested against the allometric relationship of Pefkochori. It was shown that the As values produced by the application of the two allometric equations were closely related ($R^2 = 0.996$, $p < 0.05$; Figure S1).

2.4. Radial Profile Sap Flow Monitoring

Xylem sap flux density was monitored at the five (5) Aleppo pines using the thermal dissipation method [42,43]. We installed on each of the five trees three Granier-type sensors at different sapwood depths, at 21, 51 and 81 mm. The 21 mm sapwood depth was selected as representative of the sapwood depth range most commonly used for sap flow measurements in forest trees, while the 51 mm and 81 mm sapwood depths were chosen for assessing sap flux patterns at greater depths. The studied Aleppo pines are characterized by high DBH, quite narrow heartwood, and deep sapwood (Table 1). Limited heartwood has similarly been reported in a study of *P. canariensis* [44]. Thus, we could safely assume that all sensors including the deeper ones at 81 mm were installed in conductive sapwood.

Both 51 and 81 mm sensors were developed using long gauge needles that were cut at the desirable length (51 mm and 81 mm, accordingly). In these probes the thermocouple was placed at a distance 10 mm short of the total needle length and only 20 mm of the needle tip was covered with heating wire. Each sensor was radially inserted in the outer sapwood at breast height (DBH) after the bark was removed. The reference probe was inserted approximately 12.0 cm below the heated needle in a vertical alignment. All three sensors were installed at the same height with a 120° distance among them. To exclude azimuthal effects, the three different depth sensors were established on the tree periphery the following way: the first 21 mm long sensor was installed on the north-facing side of Tree 1 and the 21 mm sensor on trees 2 to 5 was installed by moving by 75° on each next tree's periphery, in a clockwise manner. The 51 and 81 mm sensors were installed then accordingly by maintaining the 120° distance among them.

A CR10X datalogger connected to an AM 416 Multiplexer (CR10X Campbell Scientific, Logan, UT, USA) was used to heat the sensors and collect the probe outputs. The temperature difference between the Granier-type probes (ΔT) was recorded at 10 s intervals and data-logged as 15 min averages. ΔT values were related to the daily maximum temperature difference among the probes (ΔTm) to calculate sap flux density (Js, kg m^{-2} h^{-1}) according to Granier's equation [43]

$$Js = 428.4 \left(\frac{\Delta Tm - \Delta T}{\Delta T} \right)^{1.231}$$

All probes and stems were insulated to minimize natural temperature gradients. Nevertheless, heating was disconnected in all sensors from 28 May 2019 to 6 June 2019 (10 days) to monitor the natural temperature gradients. Even though ΔT values were consistently < 5.4% of the sap flow signal, and corrections were not necessary [11], we modelled the measured natural temperature gradients using meteorological variables by means of principal components regression (PCR), for each individual sensor, to correct the temperature differences measured when the sensors were heated, as described by Graham [45] and Martínez-Vilalta et al. [46]. The corrected values were used to calculate sap flux density (Js) for the three different depths per tree (kg m^{-2} h^{-1}). To scale-up to the sap flow per conducting sapwood area (Qs; kg h^{-1}) a weighted mean was calculated according to Hatton et al. [47] and Poyatos et al. [13] as following:

$$Q_s = J_{s,21} A_{s,21} + J_{s,51} A_{s,51} + J_{s,81} A_{s,81} + J_{s,81} A_{s,in}$$

where $J_{s,i}$ is sap flux density at each measuring depth, $A_{s,i}$ is the corresponding annulus area and $A_{s,in}$ is the conducting area beyond the influence of the last measuring point (deeper than 81 mm).

2.5. Data Analysis

Data on temperature difference among the probes as well as meteorological data were curated in R [48] using the lubridate package [49]. Days with any missing values were excluded from analysis. Continuous missing days in 2008–2009 were due to malfunctions in the power supply. For the development of mean annual or seasonal diurnal curves of sap flux density (Js) and environmental parameters, as well as for the assessment of the seasonal variation of sap flow rates (Qs) and environmental parameters, 60-min means (1st period) and 30-min means (2nd period) were built from original 15-min data, in accordance to the frequency of datalogging of micrometeorological parameters. Then data were averaged on a DOY (Day of the year) basis of each measuring period for each tree ($n = 5$).

Statistical analysis and data visualization were performed with OriginPro, version 2020b (OriginLab Corp., Northampton, MA, USA) and IBM SPSS Statistics, version 23 (IBM Corp., Armonk, NY, USA). Relationships of Js and Qs with VPD, SWC, and solar radiation were tested with linear multiple stepwise regression and with non-linear regression for $p < 0.05$. One way Anova and Tukey's test were used for detecting significant differences in sap flux density (Js) between the three sapwood depths and the different seasons at $p < 0.05$. Values of March to May, June to August, September to November, and December to February were used for building the seasonal means of spring, summer, autumn, and winter, respectively. Significant differences between the daily curves of Js at the three sapwood depths were detected with the Wilcoxon and Kendall non-parametric tests at $p < 0.05$.

3. Results

3.1. Climate Uring the Study Periods

Table 2 presents the long-term (2008–2019) and the annual and seasonal means of climatic traits of the forest stand during the two study periods. Overall, the wet period was characterized by substantially higher soil water content compared to the dry period. Rainfall and air temperature were also higher in the wet period, resulting in higher VPD values than in the dry period. Similarly, Tair and VPD were lower in the dry period, compared to the wet one, during each season. An exception was observed in regard to rainfall, which followed the same pattern in all seasons except from winter; then it was higher in the dry period than in the wet one. Finally, a pronounced difference was observed in SWC, which was quite lower in the summer and autumn of the dry period vs. the wet period.

Table 2. Means of air temperature (Tair), vapor pressure deficit (VPD), soil water content (SWC) and sum of rainfall at the forest stand during (**a**) the period 2008–2019, (**b**) the two study periods (Dry period: July 2008–November 2009, Wet period: April 2018–April 2020) and (**c**) the different seasons within each study period.

	Tair (°C)	VPD (KPa)	Rainfall (mm)	SWC (%)
2008–2019	16.3	0.56	604.8	18.9 [1]
Dry period	15.9	0.53	644.2	18.8
Spring	14.1	0.42	115.8	24.7
Summer	24.3	0.96	105.0	13.5
Autumn	16.7	0.47	131.0	11.7
Winter	8.7	0.28	292.4	25.1
Wet period	17.9	0.70	889.3	24.3
Spring	17.0	0.54	171.0	23.2
Summer	26.0	1.21	260.1	21.8
Autumn	19.4	0.66	299.5	22.3
Winter	9.2	0.37	158.7	29.9

[1] SWC was not recorded during July–August 2008 and March 2016–November 2017 due to malfunction.

3.2. Diurnal and Seasonal Variation of Js with Increasing Sapwood Depth

The diurnal curves of VPD and Js in the three sapwood depths on a seasonal basis are presented in Figure 1, while the seasonal means of Js are shown in Table 3 for the different sapwood depths and both study periods. The patterns and the range of Js differed when observed separately per season of the year and study period (dry vs. wet) (Figure 1).

Js at the outer sapwood (21 mm) was constantly higher than the other two depths, but the difference among depths was less pronounced in autumn, when in general the lowest Js values were recorded during both the dry and the wet period, despite the substantially higher VPD values in the latter case (Figure 1E,F, Table 3). In all other seasons, Js in 21 mm was significantly higher than in 81 mm in the dry period (Figure 1A,C,G) and significantly higher than both 51 and 81 mm in the wet period (Figure 1B,D,H). In both periods, the greater contribution of outer (21 mm) vs. inner xylem to sap flux was most pronounced in winter (41.9% difference compared to Js in 81 mm in the dry period and 31.4% difference compared to Js in 51 mm in the wet season) (Table 3).

In the dry study period, differences in the peak of Js were recorded among the seasons. Js peaked later during the day in summer and spring than in autumn and winter. No such response was observed in the wet period (Figure 1).

During the dry period, the highest Js in all depths was recorded in spring, while in the wet period Js in all tested depths was in the same range during spring and summer (Figure 1A, Table 3), probably as a result of the high summer soil water availability and evaporative demand (high air temperature and VPD; Table 2). It is also noteworthy that although Js in all sapwood depths was restricted to a shorter day phase during the winter, still the peak values of winter Js in 21 mm were comparable to those of spring and summer, both during the dry and the wet period (Figure 1).

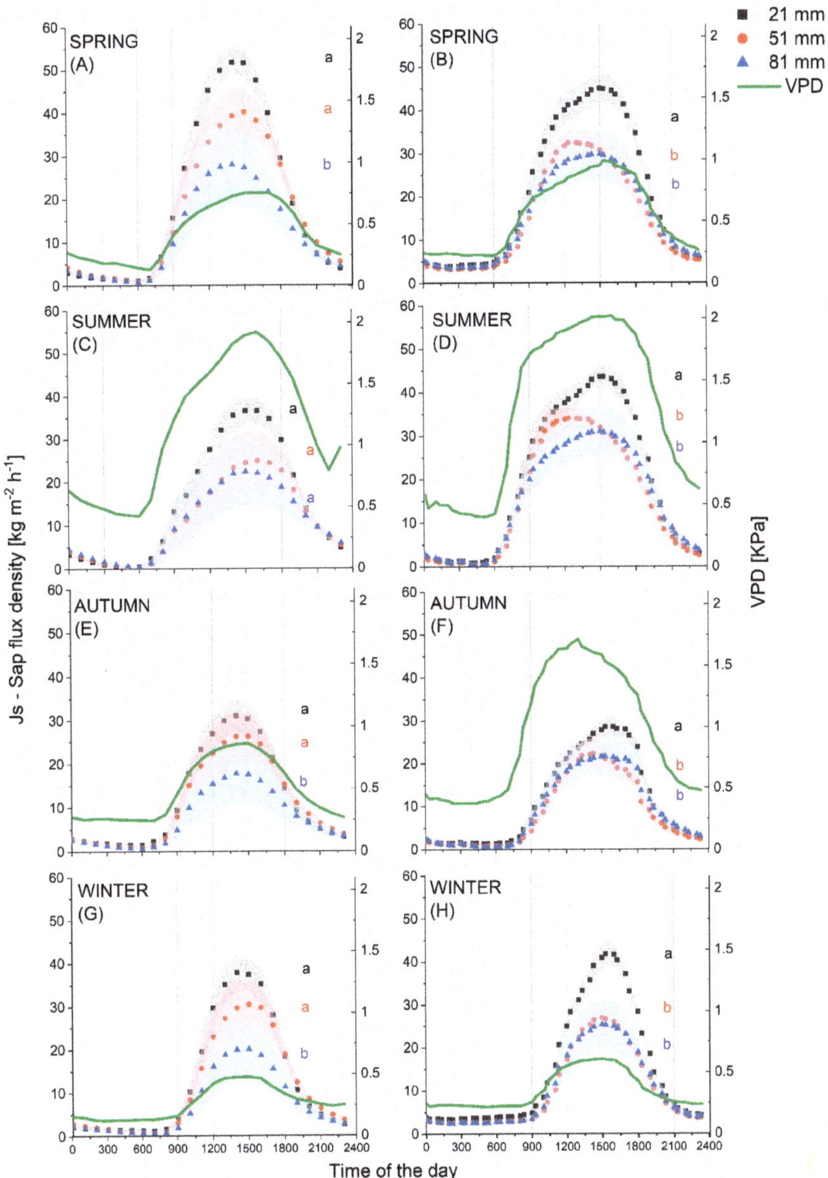

Figure 1. Diurnal variation of Js in the 3 radial depths and VPD in the different seasons of the year during the two study periods (dry period: **A,C,E,G**), wet period: **B,D,F,H**). Different letters indicate significant differences between the daily curves of Js in the 3 radial depths at $p < 0.05$ within each season and period. Symbols represent mean values ($n = 5$ trees) and the shaded bands represent $\pm$ SE of the means.

Table 3. Mean diurnal Js ± SE in the 3 different sapwood depths and in the different seasons during the two study periods (Dry period: July 2008–November 2009; wet period: April 2018–April 2020). The difference (%) between Js at 21 mm and Js at the two inner depths is also shown. Lowercase different letters indicate significant differences between the 3 sapwood depths, within each season, while uppercase different letters indicate significant differences among the seasons.

	Mean Js (kg m^{-2} h^{-1})			% Difference of Js in 21 mm Compared to	
	Dry Period				
	Js 21 mm	**Js 51 mm**	**Js 81 mm**	**Js 51 mm**	**Js 81 mm**
Spring	19.4 ± 4.0 *aA*	16.3 ± 2.9 *abA*	11.7 ± 2.1 *bA*	15.6	39.7
Summer	13.4 ± 2.8 *aAB*	11.0 ± 1.9 *aB*	10.5 ± 1.6 *aA*	25.9	30.0
Autumn	11.7 ± 2.3 *aB*	10.6 ± 1.9 *aB*	7.4 ± 1.3 *bB*	9.1	36.8
Winter	12.5 ± 2.8 *aAB*	11.0 ± 2.2 *aB*	7.3 ± 1.5 *bB*	12.1	41.9
	Wet Period				
	Js 21 mm	**Js 51 mm**	**Js 81 mm**	**Js 51 mm**	**Js 81 mm**
Spring	20.5 ± 1.7 *aA*	14.8 ± 1.5 *bA*	15.1 ± 5.7 *bA*	27.5	26.2
Summer	19.2 ± 1.6 *aA*	15.2 ± 1.6 *bA*	15.0 ± 6.0 *bA*	21.2	22.1
Autumn	11.0 ± 1.1 *aC*	8.3 ± 1.3 *bB*	9.1 ± 4.6 *abB*	24.2	17.5
Winter	14.6 ± 1.0 *aB*	10.0 ± 0.9 *bB*	10.1 ± 5.8 *abB*	31.4	30.9

The combined effect of VPD with solar radiation, or all climatic parameters, generally produced the strongest regression models describing the diurnal variation in Js during the seasons and the two study periods (Table S1). Among the tested climatic factors, VPD had the strongest control on the diurnal variation of Js in all sapwood depths (R_{adj}^2 ranged from 0.802 to 0.990). Furthermore, the effect of VPD tended to increase with increasing sapwood depth, but this pattern was not constant in all seasons. On the contrary, SWC had either the lowest or no effect on the diurnal variation of Js (Table S1).

Mean Js averaged over each study period, and its declining pattern with increasing sapwood depth is shown in Figure 2. In the dry period, Js was 13.9 ± 1.2, 11.6 ± 2.4, and 9.1 ± 3.0 kg m^{-2} h^{-1} in 21, 51, and 81 mm depth, respectively (Figure 2A), while in the wet period, the respective values were 16.6 ± 1.1, 12.2 ± 1.2, and 12.2 ± 2.4 kg m^{-2} h^{-1} (Figure 2B). In general, Js was not affected by differences in DBH or its increment between the two study periods, apart from Js in 51 mm, which increased with increasing DBH (Figure S3).

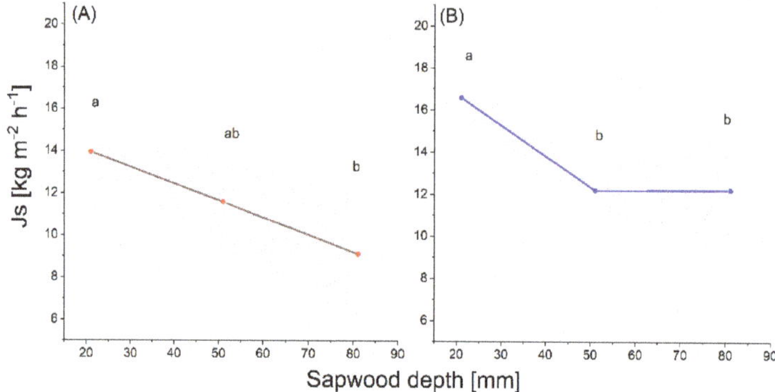

Figure 2. Variation of mean tree sap flux density (Js) with increasing radial sapwood depth during (**A**) the dry study period, and (**B**) the wet study period. Shaded bands represent ± SE of the mean (*n* = 5 trees). Different letters indicate significant differences between the 3 sapwood depths.

3.3. Tree Sap Flow Rate on a Sapwood Area Basis (Qs) and Environmental Control on Qs

In order to upscale the Js measurements to the tree level and calculate the sap flow on a sapwood area basis (Qs), a weighted mean was calculated according to Hatton et al. [47] and Poyatos et al. [13]. Considering solely Js measurements conducted at the outer sapwood (21 mm) than in different sapwood depths would result in an overestimation of Qs both in the dry period (27.7 $\pm$ 0.2 %) and in the wet period (20.2 $\pm$ 0.4 %).

For verification purposes, although we lacked information on Js in depths smaller than 21 mm, we also calculated mean Js per tree with the Gaussian function proposed by Ford et al. [27] and Martínez–Vilalta et al. [46], as described in detail in Figure S2. Strong correlations between Qs values calculated both ways were found for both study periods ($R^2 = 0.99$, $p < 0.05$, Figure S2).

The seasonal fluctuation of Qs, SWC, and VPD during the year is presented in Figure 3 for both study periods. The substantially higher SWC and VPD levels of the wet period (Figure 3D) resulted in higher sap flow rates, particularly during the mid-summer period (DOY 180–210).

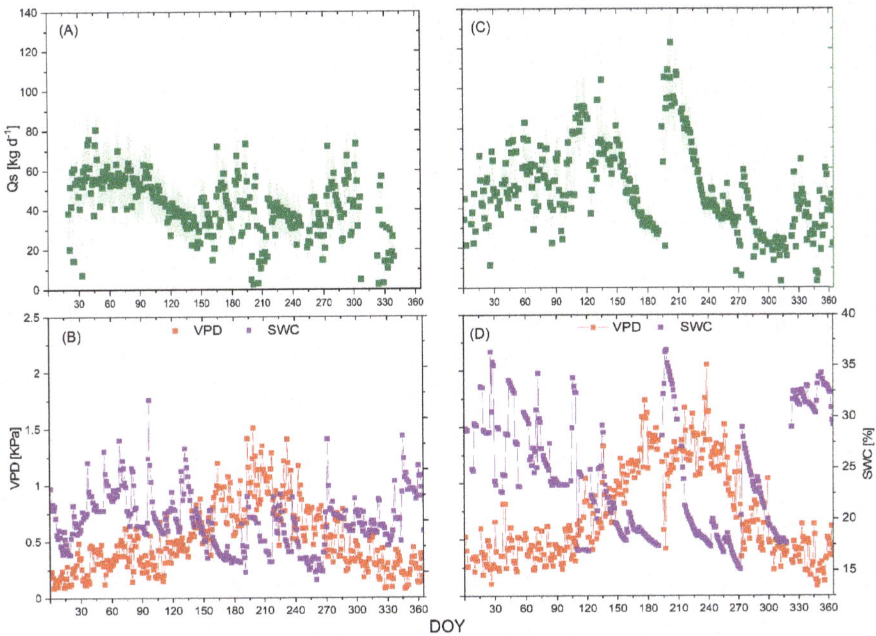

Figure 3. Seasonal variation of sap flow rate per sapwood area (Qs; kg d^{-1}), VPD, and SWC during the dry study period (**A**,**B**) and wet study period (**C**,**D**). In plots (**A**,**C**) symbols represent mean values of daily Qs and shaded bands represent $\pm$ SE (*n* = 5 trees).

Among the tested relationships between environmental parameters (VPD, SWC, and solar radiation) and Qs only solar radiation had a significant effect when all days were considered simultaneously (Table S2). The effect of climate was also tested within different thresholds of VPD and SWC (data not shown), and it was found that additional and stronger climatic controls on Qs were detected when VPD and SWC were higher than 0.7 KPa and 20%, respectively (Table S2).

In particular, solar radiation largely controlled Qs when SWC was above 20%, and this effect was stronger in the wet (R^2_{adj}: 0.579, $p < 0.001$) than in the dry period (R^2_{adj}: 0.220, $p < 0.001$) (Table S2) and better described by an allometric relationship (Figure 4A,B).

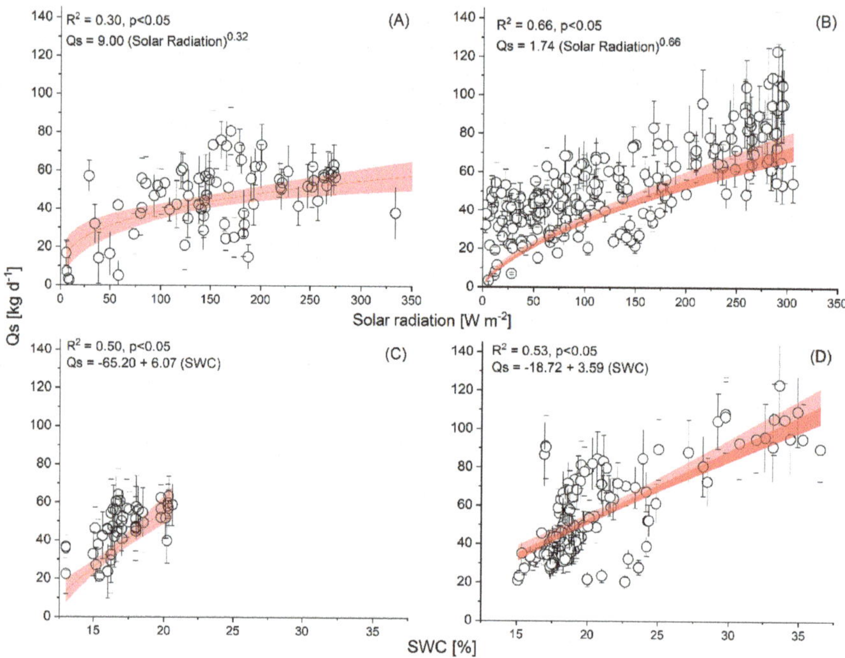

Figure 4. Relationships between Qs [kg d^{-1}] and climatic parameters during the dry period (**left**) and the wet period (**right**) are shown by the regression models between Qs and solar radiation when SWC > 20% (**A,B**) and SWC when VPD > 0.7 KPa (**C,D**). Symbols indicate mean values, error bars indicate ± SE (*n* = 5 trees) and shaded bands indicate confidence level of the regression models at *p* < 0.05.

Furthermore, a strong influence of SWC on Qs was detected in days of high evaporative demand, when VPD was above 0.7 KPa (Figure 4C,D). This effect of SWC was slightly higher in the wet period, than in the dry one, when combined with the effect of solar radiation (R^2$_{adj}$: 0.605, *p* < 0.001; Table S2).

4. Discussion

In order to more accurately estimate the sap flow of adult Aleppo pines in a natural near-coastal forest in northern Greece, we assessed the radial variability of sap flux density (Js) during two periods (July 2008–November 2009 and April 2019–April 2020), differing in climatic conditions and predominantly in water availability. Three sapwood depths were used for this purpose; (a) 21 mm as the sapwood depth range most commonly used for sap flow measurements in forest trees (e.g., [41,50,51]) and (b) 51 and 81 mm as the studied pines are of high DBH and sapwood radius (Table 1). The latter sapwood depths were safely regarded as conductive and actively contributing to the trees' sap flow. Furthermore, the variation in both sap flux density (Js) and sap flow on a sapwood area basis (Qs) is discussed in relation to temporal and climatic variability, to enhance our understanding of Aleppo pine's water balance in eastern Mediterranean ecosystems.

4.1. Sap Flux Density Changes with Increasing Sapwood Depth in Aleppo Pine

Our results indicate a considerable change in Js with increasing sapwood depth, with the highest sap flux density occurring at the outer sapwood (21 mm) of the studied Aleppo pine trees during all seasons (Figures 1 and 2), in line with studies on other conifers [29,30]. A decreasing trend in Js was presented in deeper sapwood layers (51 and 81 mm; Figures 1 and 2), which was more pronounced under drier conditions (1st study period) than under high water availability (2nd study period), when it remained stable after the depth of 51 mm (Figure 2). Similarly, Cohen et al. [35] observed an almost linear decrease of sap flux from c. 20 to 40 mm under quite arid conditions, in Aleppo pine stands in Israel. However, in this study [35], the only to our knowledge that dealt with the radial gradient of Js in *P. halepensis*, no sapwood depths greater than 44 mm were tested. Thus, no previous records are available on Js variability in Aleppo pine sapwood depths of 51 mm and 81 mm addressed in the present work.

Čermak and Nadezhdina [20], who examined the radial patterns of sap flow rate in five coniferous and four broad-leaved species also addressed the effect of water availability on the radial profile of sap flow. They found that undersaturated water conditions, xylem water content peaked at c. 20–30 mm sapwood depth, whereas under drought, it was maximized at 0–12 mm beneath the cambium. Moreover, Ganthaler et al. [52] observed in conifers that electrical resistivity increased in deeper sapwood areas, even more so as tree dehydration progressed. Such changes under different water availability conditions presumably explain the steeper radial Js patterns, we observed during the drier study period.

A linear decline in sap flux to even greater depths (from 20 to 90 mm) was reported for *Pinus brutia* in Cyprus [30]. Consistently, in maritime pine Delzon et al. [22] reported a sharp decrease in sap flow with increasing sapwood depth in trees with DBH > 42 cm. Alvarado-Barrientos et al. [53] concluded that in *Pinus patula* in Mexico the peak of Js was recorded within the outermost 20–33% of sapwood. Similar declining patterns of sap flux density or specific conductivity of the xylem with increasing sapwood depth are reported for other conifers as well, suggesting that decreasing sap flux density in the inner sapwood of mature trees is associated with declining sapwood hydraulic conductivity, due to the presence of non-functional or blocked tracheids [27,33,54,55].

Instead of a linear decrease with sapwood depth, others concluded that a curved radial profile is characterizing the sap flow pattern of conifers, which can be described by a Gaussian function [18,27,29,46]. In these cases, sap flux density increased gradually from directly below the cambium to 10–20 mm sapwood depth and decreased thereafter. However, no such pattern of radial profile in sap flow has been described in Aleppo pine [35] or in the closely related Brutian pine [30]. The lack of sap flow measurements at depths smaller than 20 mm in our study does not allow drawing any conclusions on a possible curved fit of sap flux radial profile in the studied Aleppo pines.

Nevertheless, independent of the pattern of sap flux decline with increasing sapwood depth, not taking into account the radial variability of sap flux density will result in a noticeable overestimation of sap flow rate at the tree level. In our case, if only Js at the outer sapwood would be considered, Qs of Aleppo pine would be overestimated by c. $27.7 \pm 0.2\%$ under drier conditions and by c. $20.2 \pm 0.4\%$ under higher water availability. Overestimations, based on the same rationale, were also estimated by other studies on conifers and ranged from 5.6% [29] to 250% [23], depending on temporal variation and climatic conditions.

4.2. Short-and Long-Term Temporal Differences in Sap Flow Gradient in Relation to Climate

During the dry period, Js peaked later in summer, compared to other seasons, probably in response to midday depression. Js diurnal fluctuation in outer sapwood (21 mm) and in the innermost sapwood (81 mm) were quite synchronized during the day in all seasons (Figure 1). Ford et al. [27] observed that water in inner sapwood is mobilized later in the day, when stem water gets depleted, despite possible high hydraulic resistance. In the dry period, Js in 51 mm responded in a similar way, but no such pattern was observed in the wet period for any of the deeper sapwood layers (Figure 1).

In the drier period, there was a clear gradient in sap flux density with increasing sapwood depth, particularly in spring and winter when the highest Js levels were recorded (Figure 1A,G). On the contrary, in the wet period, the contribution of outer xylem was clearly the highest, while the sapwood at 51 mm and 81 mm contributed almost equally to total sap flow in all seasons (Figure 1B,D,F,H). Similarly, Phillips et al. [33] and Nadezhdina et al. [26] reported that the outer xylem conducted considerably more water than the inner xylem under conditions associated with high Js values. Still, it cannot be overlooked that the outer sapwood had the highest sap flux densities, compared to inner sapwood, both under drier and wetter conditions (Figure 2), even during autumn when the lowest Js levels were recorded. However, the proportional contribution of Js at inner xylem to total sap flow increased during autumn (Table 3) in line with the conclusion of Ford et al. [27] that as the radial gradient becomes less steep, the inner xylem contributes more to total tree sap flow.

On a daily basis, the changes in the radial profile of Js in Aleppo pine were strongly controlled by the concurrent variation in VPD (R_{adj}^2 = 0.802–0.980 in all sapwood depths and seasons and in both study periods; Table S1). The relationship of Js to VPD tended to increase with increasing sapwood depth, but this pattern was not consistent in all seasons and study periods (Table S1). Cohen et al. [35] also reported a close control of VPD over sap flux in Aleppo pine, in accordance with studies in other conifers [26,53,56]. Such a determining role of VPD on Js during the day indicated that stomata optimized the time and extent of carbon uptake with respect to VPD, to maintain an optimal water balance. On the contrary, no effect of soil water content on the daily fluctuation of sap flux density was detected in any radial depth and season (Table S1), implying that Js responds rapidly to short-term changes in atmospheric water deficits, but not to soil water content, which is not modified as quickly during the day.

On an annual scale, Qs seemed to respond to variations in soil water content in both study periods (Figure 3), although the SWC of the dry period was substantially lower than of the wet period (Table 2), but no relationship between SWC and Qs could be established for the entire study periods (Table S2). The significant positive effect of SWC on Qs was detected only in days of high evaporative demand, when VPD was higher than 0.7 KPa (Table S2, Figure 4C,D), contrary to the negative effect of rainfall and increased Tair on sap flux density of silver fir reported by Magh et al. [57]. In line with our results, Sánchez–Costa et al. [31] observed that the sap flow of Aleppo pine in Spain responded positively to increasing soil water only on days with VPD higher than 0.4 KPa.

The controlling role of SWC is highlighted by the fact that increasing solar radiation or the combined effect of increasing solar radiation and VPD led to enhanced Qs only at SWC > 20% (Figure 4A,B, Table S2), particularly during the wet period (R_{adj}^2 = 0.557, $p < 0.001$), thus when higher soil water availability allowed stomatal water loss in favor of assimilation. This response points to a limited control of solar radiation on Qs in drier days as stomatal control aimed at minimizing water losses. A decreased sensitivity of pines ecophysiological responses to other climatic factors at low soil moisture levels has previously been reported [56,58]. Tatarinov et al. [56] argued that changes in daily transpiration were strongly dominated by soil moisture availability. Moreover, they determined a similar threshold of SWC (22%), above which increasing VPD could enhance sap flow, in a semi-arid Aleppo pine forest. Similarly, Chang et al. [29] observed a positive effect of solar radiation on sap flow only at higher, compared to lower, soil water content. The importance of adequate soil water (SWC > 20%) for Qs under both drier and wetter conditions emphasizes the role of soil water availability in the studied Aleppo pine forest, possibly due to limited access to ground water. A compact $CaCO_3$ layer observed at c. 50–150 cm soil depth at the study site, could potentially prohibit root penetration to greater depths [59].

Despite the complexity of environmental conditions within the forest stand and the difficulty in establishing firm relationships, there is a coordination of sap flow to climatic variability, and particularly to soil water availability, in the studied Aleppo pine forest.

5. Conclusions

We determined the radial variability in sap flux density of mature Aleppo pines with high DBH in a natural forest in northern Greece during two periods with different water availability. In both study periods, the contribution of outer xylem was the highest. A linear decline in sap flux density with increasing xylem depth was evident during the dry period.

In the wet period, sap flux density decreased in 51 mm sapwood depth but remained stable in inner depth. These patterns of Js radial profile were, with some variations, present in all seasons of the year. We also estimated that not taking into account the radial variability of sap flux density when aiming at upscaling to sap flow on a tree level, would lead to an overestimation of c. 20.2–27.7%, under varying soil water availability. Furthermore, our results support that temporal variations in the radial profile of Js and in sap flow rates in the studied Aleppo pine forest are associated with dynamic environmental changes. On a diurnal basis, VPD was the strongest determinant of sap flux density in all sapwood depths and in both study periods, pointing to a close stomatal regulation to short-term changes in evaporative demand, while on a larger temporal scale, SWC controlled the effects of other environmental parameters. Only when SWC was high enough (>20%) did sap flow respond positively to increasing solar radiation and VPD, indicating the decisive role of soil water availability in the studied ecosystem. These findings can broaden our knowledge on the water balance strategies of Aleppo pine, at its eastern distribution in the Mediterranean basin, where information is limited to date to regions that are more arid.

Supplementary Materials: The following are available online at https://www.mdpi.com/article/10.3390/f12010002/s1, Figure S1: Linear relationship between the sapwood depths of the five (5) studied trees, calculated with the two allometric equations developed from Aleppo pine trees at Pefkochori and Sani, Chalkidiki, Greece. The shaded band represents the confidence level of the regression fit at p <0.001, Figure S2: Relationship between the sap flow rates on a sapwood area basis calculated as weighted means [13,47] (Qs, M) of the 3 sapwood depths and by the Gaussian equation (Qs, G) of Ford et al. [27] and Martínez–Vilalta et al. [46] for the dry (A) and the wet (B) period, Figure S3: Relationship between sap flux density (Js) at the three studied sapwood depths and diameter at breast height (DBH). Symbols represent mean values of Js, averaged per DOY and tree (n = 365), corresponding to the DBH of the studied Aleppo pines during the two study periods (Dry period: 1 July 2008–30 November 2009; Wet period: 13 April 2018–13 April 2020). The regression line indicates the significant relationship between Js and DBH which was found only for Js at 51 mm sapwood depth, Table S1: Multiple stepwise regression analysis of the relationships between the diurnal variation of sap flux density (Js) at the 3 sapwood depths (21, 51 and 81 mm) and climatic parameters (vapor pressure deficit-VPD, soil water content-SWC and solar radiation–Rad) during the different seasons (spring, summer, autumn, winter) of the two study periods (Dry period: 1 July 2008–30 November 2009; Wet period: 13 April 2018–13 April 2020). The adjusted R^2 and p level of each regression model are presented. The models with the highest R^2 and significance level are indicated with bold, while non-significant models are indicated with n.s., Table S2: Multiple stepwise regression analysis of the relationships between mean weighted sap flow on a sapwood area basis (Qs) and climatic parameters (vapor pressure deficit-VPD, soil water content-SWC and solar radiation–Rad) during the two study periods (Dry period: 1 July 2008–30 November 2009; Wet period: 13 April 2018–13 April 2020). Regression models were tested including all days of each study period (whole period), only days with VPD > 0.7 KPa or only days with SWC > 20%. The adjusted R^2 and p level of each regression model are presented. The models with the highest R^2 and significance level are indicated with bold, while non-significant models are indicated with n.s.

Author Contributions: Conceptualization, M.N.F.; methodology, E.K. and M.N.F.; formal analysis, E.K. and M.N.F.; investigation, M.N.F.; resources, M.N.F.; data curation, E.K. and M.N.F.; writing—original draft preparation, E.K.; writing—review and editing, M.N.F.; visualization, M.N.F. Both authors have read and agreed to the published version of the manuscript.

Funding: This research received no external funding.

Institutional Review Board Statement: Not applicable.

Informed Consent Statement: Not applicable.

Data Availability Statement: The data presented in this study are available from the corresponding author upon reasonable request.

Acknowledgments: N. Fyllas, Aegean University, is kindly acknowledged for his assistance on data curation with R. We also wish to thank G. Spyroglou for the biometrical characterization of the studied trees and M. Dimitriou for his precious technical support during sap flow sensors preparation and installation. We are grateful to K. Radoglou, Democritus University of Thrace, and Sani Resort S.A. for the provision of datalogging and power supply infrastructure at the study site. The authors wish to thank the anonymous reviewer for the helpful comments and suggestions during the revision stage.

Conflicts of Interest: The authors declare no conflict of interest.

References

1. Granier, A.; Huc, R.; Barigah, S. Transpiration of natural rain forest and its dependence on climatic factors. *Agric. For. Meteorol.* **1996**, *78*, 19–29. [CrossRef]
2. Biron, P.; Siegwolf, R.; Granier, A. Estimates of water vapor flux and canopy conductance of Scots pine at the tree level utilizing different xylem sap flow methods. *Theor. Appl. Clim.* **1996**, *53*, 105–113. [CrossRef]
3. Pataki, D.E.; Oren, R.; Phillips, N. Responses of sap flux and stomatal conductance of Pinus taeda L. trees to stepwise reductions in leaf area. *J. Exp. Bot.* **1998**, *49*, 871–878. [CrossRef]
4. Tsuruta, K.; Komatsu, H.; Kume, T.; Shinohara, Y.; Otsuki, K. Canopy transpiration in two Japanese cypress forests with contrasting structures. *J. For. Res.* **2015**, *20*, 464–474. [CrossRef]
5. Oren, R.; Phillips, N.; Ewers, B.E.; Pataki, D.E.; Megonigal, J.P. Sap-flux-scaled transpiration responses to light, vapor pressure deficit, and leaf area reduction in a flooded Taxodium distichum forest. *Tree Physiol.* **1999**, *19*, 337–347. [CrossRef]
6. Ryan, M.G.; Bond, B.J.; Law, B.E.; Hubbard, R.M.; Woodruff, D.; Cienciala, E.; Kucera, J. Transpiration and whole-tree conductance in ponderosa pine trees of different heights. *Oecologia* **2000**, *124*, 553–560. [CrossRef]
7. Köstner, B.M.M.; Schulze, E.-D.; Kelliher, F.M.; Hollinger, D.Y.; Byers, J.N.; Hunt, J.E.; McSeveny, T.M.; Meserth, R.; Weir, P.L. Transpiration and canopy conductance in a pristine broad-leaved forest of Nothofagus: An analysis of xylem sap flow and eddy correlation measurements. *Oecologia* **1992**, *91*, 350–359. [CrossRef]
8. Granier, A.; Loustau, D. Measuring and modelling the transpiration of a maritime pine canopy from sap-flow data. *Agric. For. Meteorol.* **1994**, *71*, 61–81. [CrossRef]
9. Oren, R.; Phillips, N.G.; Katul, G.; Ewers, B.E.; Pataki, D.E. Scaling xylem sap flux and soil water balance and calculating variance: A method for partitioning water flux in forests. *Annal. Sci. For.* **1998**, *55*, 191–216. [CrossRef]
10. Ducrey, M.; Duhoux, F.; Huc, R.; Rigolot, E. The ecophysiological and growth responses of Aleppo pine (Pinushalepensis) to controlled heating applied to the base of the trunk. *Can. J. For. Res.* **1996**, *26*, 1366–1374. [CrossRef]
11. Do, F.; Rocheteau, A. Influence of natural temperature gradients on measurements of xylem sap flow with thermal dissipation probes. 2. Advantages and calibration of a noncontinuous heating system. *Tree Physiol.* **2002**, *22*, 649–654. [CrossRef] [PubMed]
12. Martínez-Vilalta, J.; Mangirón, M.; Ogaya, R.; Sauret, M.; Serrano, L.; Peñuelas, J.; Piñol, J. Sap flow of three co-occurring Mediterranean woody species under varying atmospheric and soil water conditions. *Tree Physiol.* **2003**, *23*, 747–758. [CrossRef] [PubMed]
13. Poyatos, R.; Llorens, P.; Gallart, F. Transpiration of montane Pinus sylvestris L. and Quercus pubescens Willd. forest stands measured with sap flow sensors in NE Spain. *Hydrol. Earth Syst. Sci. Discuss.* **2005**, *2*, 1011–1046. [CrossRef]
14. Renninger, H.J.; Clark, K.L.; Skowronski, N.S.; Schäfer, K.V.R. Effects of a prescribed fire on water use and photosynthetic capacity of pitch pines. *Trees* **2013**, *27*, 1115–1127. [CrossRef]
15. Sánchez-Costa, E.; Sabaté, S.; Poyatos, R. Identifying water use strategies in mediterranean tree species combining sap flow and dendrometer data. *Acta Hortic.* **2013**, 269–275. [CrossRef]
16. Mitchell, P.J.; O'Grady, A.P.; Tissue, D.T.; Worledge, D.; Pinkard, E.A. Co-ordination of growth, gas exchange and hydraulics define the carbon safety margin in tree species with contrasting drought strategies. *Tree Physiol.* **2014**, *34*, 443–458. [CrossRef]
17. Zhu, L.; Wang, Q.; Ni, G.; Niu, J.; Zhao, P.; Zhang, Z.; Zhao, P.; Gao, J.; Huang, Y.; Gu, D.; et al. Stomatal and hydraulic conductance and water use in a eucalypt plantation in Guangxi, southern China. *Agric. For. Meteorol.* **2015**, *202*, 61–68. [CrossRef]
18. Berdanier, A.B.; Miniat, C.F.; Clark, J.S. Predictive models for radial sap flux variation in coniferous, diffuse-porous and ring-porous temperate trees. *Tree Physiol.* **2016**, *36*, 932–941. [CrossRef]
19. Čermák, J.; Cienciala, E.; Kučera, J.; Hällgren, J.-E. Radial velocity profiles of water flow in trunks of Norway spruce and oak and the response of spruce to severing. *Tree Physiol.* **1992**, *10*, 367–380. [CrossRef]
20. Čermák, J.; Nadezhdina, N. Sapwood as the scaling parameter- defining according to xylem water content or radial pattern of sap flow? *Annal. Sci. For.* **1998**, *55*, 509–521. [CrossRef]
21. Lu, P.; Müller, W.J.; Chacko, E.K. Spatial variations in xylem sap flux density in the trunk of orchard-grown, mature mango trees under changing soil water conditions. *Tree Physiol.* **2000**, *20*, 683–692. [CrossRef]
22. Delzon, S.; Sartore, M.; Granier, A.; Loustau, D. Radial profiles of sap flow with increasing tree size in maritime pine. *Tree Physiol.* **2004**, *24*, 1285–1293. [CrossRef] [PubMed]
23. Fiora, A.; Cescatti, A. Diurnal and seasonal variability in radial distribution of sap flux density: Implications for estimating stand transpiration. *Tree Physiol.* **2006**, *26*, 1217–1225. [CrossRef] [PubMed]

24. Reyes-Acosta, J.L.; Lubczynski, M.W. Optimization of dry-season sap flow measurements in an oak semi-arid open woodland in Spain. *Ecohydrology* **2012**, *7*, 258–277. [CrossRef]

25. Zhang, J.-G.; He, Q.-Y.; Shi, W.-Y.; Otsuki, K.; Yamanaka, N.; Du, S.; Du, S. Radial variations in xylem sap flow and their effect on whole-tree water use estimates. *Hydrol. Process.* **2015**, *29*, 4993–5002. [CrossRef]

26. Nadezhdina, N.; Cermák, J.; Ceulemans, R.J. Radial patterns of sap flow in woody stems of dominant and understory species: Scaling errors associated with positioning of sensors. *Tree Physiol.* **2002**, *22*, 907–918. [CrossRef]

27. Ford, C.R.; McGuire, M.A.; Mitchell, R.J.; Teskey, R.O. Assessing variation in the radial profile of sap flux density in Pinus species and its effect on daily water use. *Tree Physiol.* **2004**, *24*, 241–249. [CrossRef]

28. Krauss, K.W.; Duberstein, J.A. Sapflow and water use of freshwater wetland trees exposed to saltwater incursion in a tidally influenced South Carolina watershed. *Can. J. For. Res.* **2010**, *40*, 525–535. [CrossRef]

29. Chang, X.; Zhao, W.; He, Z. Radial pattern of sap flow and response to microclimate and soil moisture in Qinghai spruce (Picea crassifolia) in the upper Heihe River Basin of arid northwestern China. *Agric. For. Meteorol.* **2014**, *187*, 14–21. [CrossRef]

30. Eliades, M.; Bruggeman, A.; Djuma, H.; Lubczynski, M.W. Tree Water Dynamics in a Semi-Arid, Pinus brutia Forest. *Water* **2018**, *10*, 1039. [CrossRef]

31. Sánchez-Costa, E.; Poyatos, R.; Sabaté, S. Contrasting growth and water use strategies in four co-occurring Mediterranean tree species revealed by concurrent measurements of sap flow and stem diameter variations. *Agric. For. Meteorol.* **2015**, *207*, 24–37. [CrossRef]

32. Bush, S.; Hultine, K.R.; Sperry, J.S.; Ehleringer, J.R. Calibration of thermal dissipation sap flow probes for ring- and diffuse-porous trees. *Tree Physiol.* **2010**, *30*, 1545–1554. [CrossRef]

33. Phillips, N.; Oren, R.; Zimmermann, R. Radial patterns of xylem sap flow in non-, diffuse- and ring-porous tree species. *Plant. Cell Environ.* **1996**, *19*, 983–990. [CrossRef]

34. Hernandez-Santana, V.; Fernández, J.; Rodriguez-Dominguez, C.; Romero, R.; Diaz-Espejo, A. The dynamics of radial sap flux density reflects changes in stomatal conductance in response to soil and air water deficit. *Agric. For. Meteorol.* **2016**, *2016*, 92–101. [CrossRef]

35. Cohen, Y.; Cohen, S.; Cantuarias-Avilés, T.; Schiller, G. Variations in the radial gradient of sap velocity in trunks of forest and fruit trees. *Plant. Soil* **2007**, *305*, 49–59. [CrossRef]

36. Euro+Med (2006+). Euro+Med PlantBase—The Information Resource for Euro-Mediterranean Plant Diversity. Available online: http://ww2.bgbm.org/EuroPlusMed/ (accessed on 20 June 2020).

37. Chambel, R.; Climent, J.; Pichot, C.; Ducci, F. Mediterranean Pines (Pinus halepensis Mill. and brutia Ten.). In *Managing Forest Ecosystems*; Springer Science and Business Media LLC: Berlin, Germany, 2013; Volume 25, pp. 229–265.

38. García-Ruiz, J.M.; López-Moreno, J.I.; Vicente-Serrano, S.M.; Lasanta–Martínez, T.; Beguería, S. Mediterranean water resources in a global change scenario. *Earth-Sci. Rev.* **2011**, *105*, 121–139. [CrossRef]

39. Sarris, D.; Christodoulakis, D.; Körner, C. Recent decline in precipitation and tree growth in the eastern Mediterranean. *Glob. Chang. Biol.* **2007**, *13*, 1187–1200. [CrossRef]

40. Schiller, G.; Cohen, Y. Water regime of a pine forest under a Mediterranean climate. *Agric. For. Meteorol.* **1995**, *74*, 181–193. [CrossRef]

41. Fotelli, M.; Korakaki, E.; Paparrizos, S.; Radoglou, K.; Awada, T.; Matzarakis, A. Environmental Controls on the Seasonal Variation in Gas Exchange and Water Balance in a Near-Coastal Mediterranean *Pinus halepensis* Forest. *Forests* **2019**, *10*, 313. [CrossRef]

42. Granier, A. A new method of sap flow measurement in tree stems. *Ann. Sci. For.* **1985**, *42*, 193–200. [CrossRef]

43. Granier, A. Evaluation of transpiration in a Douglas-fir stand by means of sap flow measurements. *Tree Physiol.* **1987**, *3*, 309–320. [CrossRef] [PubMed]

44. Climent, J.; Gil, L.; Pardos, J. Heartwood and sapwood development and its relationship to growth and environment in Pinus canariensis Chr.Sm ex DC. *For. Ecol. Manage.* **1993**, *59*, 165–174. [CrossRef]

45. Graham, M.H. confronting multicollinearity in ecological multiple regression. *Ecology* **2003**, *84*, 2809–2815. [CrossRef]

46. Martínez-Vilalta, J.; Vanderklein, D.; Mencuccini, M. Tree height and age-related decline in growth in Scots pine (Pinus sylvestris L.). *Oecologia* **2006**, *150*, 529–544. [CrossRef]

47. Hatton, T.J.; Catchpole, E.A.; Vertessy, R.A. Integration of sapflow velocity to estimate plant water use. *Tree Physiol.* **1990**, *6*, 201–209. [CrossRef]

48. R Core Team. *R: A Language and Environment for Statistical Computing*; R Foundation for Statistical Computing: Vienna, Austria, 2019.

49. Grolemund, G.; Wickham, H. Dates and Times Made Easy with lubridate. *J. Stat. Softw.* **2011**, *40*, 1–25. [CrossRef]

50. Awada, T.; El-Hage, R.; Geha, M.; Wedin, D.A.; Huddle, J.A.; Zhou, X.; Msanne, J.; Sudmeyer, R.A.; Martin, D.L.; Brandle, J.R. Intra-annual variability and environmental controls over transpiration in a 58-year-old even-aged stand of invasive woodyJuniperus virginianaL. in the Nebraska Sandhills, USA. *Ecohydrology* **2012**, *6*, 731–740. [CrossRef]

51. Wang, Q.; Lintunen, A.; Zhao, P.; Shen, W.; Salmon, Y.; Chen, X.; Ouyang, L.; Zhu, L.; Ni, G.-Y.; Sun, D.; et al. Assessing Environmental Control of Sap Flux of Three Tree Species Plantations in Degraded Hilly Lands in South China. *Forests* **2020**, *11*, 206. [CrossRef]

52. Ganthaler, A.; Sailer, J.; Bär, A.; Losso, A.; Mayr, S. Noninvasive Analysis of Tree Stems by Electrical Resistivity Tomography: Unraveling the Effects of Temperature, Water Status, and Electrode Installation. *Front. Plant. Sci.* **2019**, *10*. [CrossRef]

53. Alvarado-Barrientos, M.S.; Hernandez-Santana, V.; Asbjornsen, H. Variability of the radial profile of sap velocity in Pinus patula from contrasting stands within the seasonal cloud forest zone of Veracruz, Mexico. *Agric. For. Meteorol.* **2013**, *168*, 108–119. [CrossRef]

54. Sperry, J.S.; Perry, A.H.; Sullivan, J.E.M. Pit Membrane Degradation and Air-Embolism Formation in Ageing Xylem Vessels ofPopulus tremuloidesMichx. *J. Exp. Bot.* **1991**, *42*, 1399–1406. [CrossRef]
55. Spicer, R.; Gartner, B. The effects of cambial age and position within the stem on specific conductivity in Douglas-fir (Pseudotsuga menziesii) sapwood. *Trees* **2001**, *15*, 222–229. [CrossRef]
56. Tatarinov, F.; Rotenberg, E.; Maseyk, K.; Ogée, J.; Klein, T.; Yakir, D. Resilience to seasonal heat wave episodes in a Mediterranean pine forest. *New Phytol.* **2016**, *210*, 485–496. [CrossRef] [PubMed]
57. Magh, R.-K.; Bonn, B.; Grote, R.; Burzlaff, T.; Pfautsch, S.; Rennenberg, H. Drought Superimposes the Positive Effect of Silver Fir on Water Relations of European Beech in Mature Forest Stands. *Forests* **2019**, *10*, 897. [CrossRef]
58. Addington, R.N.; Mitchell, R.J.; Oren, R.; Donovan, L.A. Stomatal sensitivity to vapor pressure deficit and its relationship to hydraulic conductance in Pinus palustris. *Tree Physiol.* **2004**, *24*, 561–569. [CrossRef] [PubMed]
59. Orfanoudakis, M.; Democritus University of Thrace, Nea Orestiada, Greece. Personal Communication, 2020.

Article

Stomatal and Leaf Morphology Response of European Beech (*Fagus sylvatica* L.) Provenances Transferred to Contrasting Climatic Conditions

Peter Petrík [1,*,†], Anja Petek [1,†], Alena Konôpková [1], Michal Bosela [1], Peter Fleischer [1,2], Josef Frýdl [3] and Daniel Kurjak [1]

1   Faculty of Forestry, Technical University in Zvolen, T. G. Masaryka 24, 960 53 Zvolen, Slovakia; anja.petek1@gmail.com (A.P.); alena.konopkova@tuzvo.sk (A.K.); ybosela@tuzvo.sk (M.B.); xfleischer@tuzvo.sk (P.F.); kurjak@tuzvo.sk (D.K.)
2   Institute of Forest Ecology, Slovak Academy of Sciences, Ľ. Štúra 2, 960 53 Zvolen, Slovakia
3   Forestry and Game Management Research Institute, Strnady 136, 252 02 Jíloviště, Czech Republic; frydl@vulhm.cz
*   Correspondence: peterpetrik94@gmail.com; Tel.: +421-915-830-543
†   Contributed equally to this work.

Received: 12 November 2020; Accepted: 16 December 2020; Published: 18 December 2020

Abstract: Climate change-induced elevated temperatures and drought are considered to be serious threats to forest ecosystems worldwide, negatively affecting tree growth and viability. We studied nine European beech (*Fagus sylvatica* L.) provenances located in two provenance trial plots with contrasting climates in Central Europe. Stomata play a vital role in the water balance of plants by regulating gaseous exchanges between plants and the atmosphere. Therefore, to explain the possible adaptation and acclimation of provenances to climate conditions, stomatal (stomatal density, the length of guard cells, and the potential conductance index) and leaf morphological traits (leaf size, leaf dry weight and specific leaf area) were assessed. The phenotypic plasticity index was calculated from the variability of provenances' stomatal and leaf traits between the provenance plots. We assessed the impact of various climatic characteristics and derived indices (e.g., ecodistance) on intraspecific differences in stomatal and leaf traits. Provenances transferred to drier and warmer conditions acclimated through a decrease in stomatal density, the length of guard cells, potential conductance index, leaf size and leaf dry weight. The reduction in stomatal density and the potential conductance index was proportional to the degree of aridity difference between the climate of origin and conditions of the new site. Moreover, we found that the climate heterogeneity and latitude of the original provenance sites influence the phenotypic plasticity of provenances. Provenances from lower latitudes and less heterogeneous climates showed higher values of phenotypic plasticity. Furthermore, we observed a positive correlation between phenotypic plasticity and mortality in the arid plot but not in the more humid plot. Based on these impacts of the climate on stomatal and leaf traits of transferred provenances, we can improve the predictions of provenance reactions for future scenarios of global climate change.

Keywords: acclimation; adaptation; common garden; drought; ecodistance; mortality; phenotypic plasticity; stomatal frequency; stomatal size

1. Introduction

European beech forests may be seriously affected by climate change-induced drought due to their well-known vulnerability to water shortages [1,2]. Combinations of heat and drought stress may cause a decrease in the vitality and competitive ability of beech populations [3–5]. There have been reports of beech populations facing strong selective pressures [6], which are foreseen to become more intense due to upcoming alterations in rainfall patterns and temperatures with ongoing climate change [7,8]. To mitigate these negative effects on the future performance of beech forests in afforestation programs in Europe, there has been increased interest in research on the intraspecific variation in beech responses to environmental changes [9–11].

Large intraspecific differences in morphological and physiological traits among the beech provenances of distinct origin reflect possible strategies which are expected to modify their response to drought. Beech populations show divergent water use strategies reflected in the differences of photosynthetic performance, water-use efficiency, leaf water potential, xylem embolism resistance and leaf morphology [10–13]. The intraspecific variation in tolerance of water deficit follows a pattern shaped by both regional and local scale effects. Beech populations originating from the sites with low precipitation [14–16], lower altitude [17] or marginal distribution range [9,18,19] show higher drought resistance in comparison with the populations from more humid environments. The observed functional variation between beech populations reaffirms the importance of local adaptation to water deficit in the context of climate change [13,20].

Common garden experiments allow us to assess the relative importance of adaptation to the site of origin and acclimation to the new environment in the expression of phenotypic traits, as all provenances are exposed to the same conditions in provenance trial plots [21]. As a result of adaptation to local original conditions, the performance and vitality of populations show a correlation with ecological characteristics at the site of origin, even after their transfer to a new environment. The effect of environmental change on a provenance planted at a given location can be expressed as the difference between the ecological characteristics of the trial plot and the site of provenance origin, called the ecodistance [22]. Moreover, if we study the performance of populations in different provenance trial plots and the differences between plots are greater than within, we expect that the differences between phenotypes are driven more by acclimation to current environmental conditions than by local adaptation [23,24]. Stomatal and leaf morphological traits such as stomatal density, potential conductance index and specific leaf area affect stomatal conductance and transpiration (functional traits) which in return influence performance, growth and survival [13,19]. Therefore, stomatal and leaf morphological traits represent a viable means to identify populations suitable for a specific environment.

The populations that possess stomatal and leaf morphological traits adapted to drought and heat stress will have an evolutionary advantage under future scenarios [3]. Hence, a plant strategy to cope with differences in water regimes involves altering stomatal density and stomatal size [25–27]. Some studies have shown that smaller stomata close more quickly than larger stomata do, thus indicating that this could enhance plant adaptation to drought [28]. However, there remains the debated issue of how stomatal density varies within a particular environment. It has been reported that drought resistant plants show higher stomatal density [27,29,30], but the results of more recent studies performed in controlled environments suggest that lower stomatal density improves drought tolerance [31,32]. Other functional traits frequently utilized in ecological studies and linked to drought tolerance are specific leaf area and leaf size. Several studies have revealed that changes in environmental factors such as light, temperature or nutrients strongly influence leaf traits [33,34]. Species with smaller, thicker leaves mainly occur in more stressful habitats and exhibit lower specific leaf areas. This trait is related to the species water use strategy [34], and it is highly plastic [35], although the precise physiological regulation mechanism of specific leaf area is still uncertain [33,36].

Furthermore, the phenotypic plasticity, defined as the capacity for a genotype to alter its morphology and/or physiology under altered environmental conditions [37], can play a major role in the survival and sustainability of forest populations subjected to global change [38–40]. This is generally seen as favourable under stress conditions because it enables plants to react to fluctuations in the environment [41,42]. However, several studies have reported a potential trade-off between phenotypic plasticity and individual fitness [43], suggesting reduced performance with increasing plasticity [38,44,45]. Therefore, it is crucial to assess the ability for stomatal and leaf morphological traits to react plastically to their environment and to test the connection between their phenotypic plasticity and plant fitness and performance.

We investigated stomatal and leaf morphological traits, their phenotypic plasticity and a link to the climate of origin and current climate in nine European beech provenances located in two provenance trial plots with contrasting climates (warmer and drier/colder and more humid) in Central Europe. Based on the premise that environmental differences between provenance trial plots can alter the stomatal and leaf morphological traits of European beech, we hypothesized that (i) provenances that grow in drier and warmer provenance plots will exhibit lower values of measured traits than those growing in more humid and colder plots to increase their performance under suboptimal conditions. We further expected to find that (ii) the climate of the provenance's original site would affect the provenance's phenotype even 18 years after transfer to a different environment. In addition to the relationship between the climate of origin and phenotype itself, we also hypothesized that (iii) the provenance climate of origin should affect the phenotypic plasticity of provenances, where provenances from more heterogeneous environments show higher phenotypic plasticity. Finally, we hypothesized (iv) a negative relationship between phenotypic plasticity and tree mortality, and provenances with higher value of plasticity would acclimate better under different environments, thus mitigating the risk of mortality.

2. Materials and Methods

2.1. Locality Description and Plant Material

The material used for this experiment was collected from two European beech (*Fagus sylvatica* L.) provenance trial plots: Tále in the Slovak Republic (near Zvolen, 48°38′ N, 19°02′ E, 810 m a.s.l.) and Zbraslav in the Czech Republic (near Prague, 49°57′ N, 14°22′ E, 360 m a.s.l.). The Slovak provenance plot included loam soil with good nutrient availability and high water holding capacity, while the Czech provenance plot included sandy loam soil with poor nutrient availability and average water holding capacity [12]. Climate data for the original provenance sites were obtained from the WorldClim high-resolution climate database [46]. The climate characteristics of the Czech provenance plot were obtained from the Praha-Libuš meteorological station, and freely available data were provided by the Czech Hydrometeorological Institute (http://portal.chmi.cz). Climate data for the Slovak provenance plot were obtained from the nearby Kremnické Bane meteorological station monitored by the Slovak Hydrometeorological Institute. We calculated additional indices from the above climate data:

Ellenberg quotient (*EQ*) [47]

$$EQ = 1000 \times \frac{T_h}{Prec} \tag{1}$$

T_h—mean temperature of the hottest month (here July), *Prec*—annual sum of precipitation.

Forest Aridity Index (*FAI*) [48]

$$FAI = 100 \times \frac{T_{7-8}}{Prec_{5-7} + Prec_{7-8}} \tag{2}$$

T_{7-8}—mean temperature of July and August, $Prec_{5-7}$—precipitation sum for May to July, $Prec_{7-8}$—precipitation sum for July to August.

Isothermality (*Iso_T*)

$$Iso_T = \frac{T \times (T_h - T_c)}{T_{MAX} - T_{MIN}} \times 100 \tag{3}$$

T—annual mean temperature, T_h—mean temperature of warmest month, T_c—mean temperature of coldest month, T_{MAX}—max temperature of warmest month, T_{MIN}—min temperature of coldest month.

Precipitation seasonality (*Seas_Prec*)

$$Seas_{Prec} = \frac{\sigma_{Prec}}{Prec} \tag{4}$$

σ_{Prec}—annual standard deviation of precipitation, *Prec*—annual sum of precipitation.

All geographical and climatic data are presented in Table 1. Optimal hydric conditions of European beech stands are represented by *EQ* values of below 20. European beech starts to lose its competitive performance in environments with *EQs* of above 20 and is replaced by more xerotic tree species in places with *EQs* of above 30 [49,50]. Moreover, locations with optimal rainfall patterns during the vegetation season for European beech are defined by an *FAI* of under 4.75 [48,51]. The *EQ* for the Czech provenance plot is 33.5, and the *FAI* is 5.1, which characterizes the plot as a location with marginal environmental conditions for European beech occurrence. The Slovak provenance plot with an *EQ* of 19.1 and *FAI* of 2.5 represents the optimal hydric environment according to the classifications mentioned above. Both provenance plots were established in 1998 as a part of the European provenance plot network, whereas proven ances were planted as two-year-old seedlings [52]. Both plots were planted with 2 × 1 m spacing under a randomized block design with three blocks and fifty seedlings per block. Nine provenances were chosen for the analysis to capture the whole altitude range of the distribution of European beech (Figure 1). The distribution of provenance original sites regarding climate characteristics is visualized in Figure A1 of the Appendix A. We were not able to sample some of the provenances in multiple blocks due to high spatial mortality. The third block of the Czech provenance plot completely died off, and some of the provenances remained only in one block. Similarly, the third block of the Slovak provenance plot suffered high mortality, and we avoided this to minimize sample heterogeneity due to unknown factors.

Table 1. Geographic and climate characteristics of provenance plots and their original locations.

Provenance	Long	Lat	Alt	T	T_{59}	Prec	$Prec_{59}$	EQ	FAI	Iso_T	$Seas_{prec}$
FR04	2.58	44.15	850	10.8	16.8	804	344	23.8	5.4	3.8	14
LUX12	6.2	49.67	400	8.6	14.9	866	365	19.7	4.8	3.1	11
UK17	−3.42	57.67	10	8.2	12.7	671	303	21.8	5.3	3.6	19
SWE23	13.2	55.57	40	7.9	14.3	640	286	25.9	6.2	2.6	21
GER26	10.67	53.65	55	8.3	15	678	319	25.5	5.1	3.0	17
AU35	14.1	47.72	1250	2.4	9.2	1495	779	7.6	1.3	3.2	26
AU36	14.85	47.53	1100	2.9	9.9	1168	648	10.4	1.6	3.1	32
PL43	22.82	49.25	900	6.3	14.1	762	433	21.5	2.9	2.7	35
PL67	18.17	54.33	250	5.8	13.2	633	336	24.6	4.3	2.4	30
CZ Zbraslav	14.37	49.95	360	8.25	15.6	532	330	33.5	5.1	na	na
SK Tále	19.03	48.63	850	6.58	14.1	842	441	19.1	2.5	na	na

Long—longitude, *Lat*—latitude, *Alt*—altitude, *T*—annual average temperature, T_{59}—average temperature during the vegetation season, *Prec*—sum of annual precipitation, $Prec_{59}$—sum of precipitation during the vegetation season, *EQ*—Ellenberg quotient, *FAI*—Forest Aridity Index, Iso_T—isothermality, $Seas_{prec}$—precipitation seasonality, na—not available.

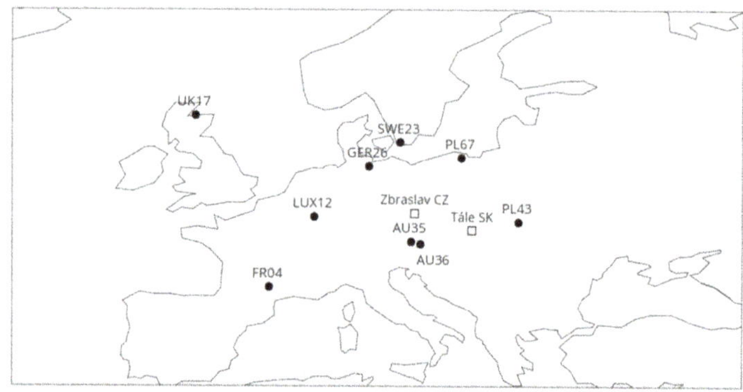

Figure 1. Localities of the tested provenances (dots) and provenance plots (squares).

2.2. Stomatal and Leaf Morphological Traits

The samples were taken during June 2016 from full sun-exposed leaves located in the upper third of the crown to minimize irradiation and canopy position effects on stomatal and leaf morphology development [53,54]. We sampled six individuals per provenance per plot and made two imprints per individual. The imprints were made by the application of transparent nail polish to the abaxial side of the leaves [55]. The layer of polish was then transferred to a microscope slide with transparent tape. To avoid possible variations in stomatal distribution within the leaves, we took imprints between the second and third veins from the base of the leaves [17]. Six images were taken from each imprint using a Motic BA210 microscope with an integrated camera (Motic Electric, Linz, Austria). Three of these images were captured at 40×10 magnification, and three photos were captured at 10×10 magnification. The images at 10×10 magnification were used to assess stomatal density (SD). The number of stomata was calculated within a 750×750 µm square per image with a random position using ImageJ 1.51k software (National Institute of Health, Bethesda, MD, USA). The assessed value of the number of stomata per square was converted to the number per square millimetre. The length of guard cells (L_A) was measured for ten stomata in a 40×10 magnification image using ImageJ software. The SD and L_A values were then averaged per individual for further analysis. Additional leaves from the same branch were scanned with a HP Scanjet G4010 scanner (Hewlett Packard, California, USA), and the leaf size (S_{leaf}) was subsequently measured by ImageJ software. The scanned leaves were stored in silica gel, after which the leaves were dried at 75 °C to a constant weight (approximately 48 h). Afterwards, the dry weight (m_{leaf}) was assessed. From the measured parameters, we calculated the following traits:

The potential conductance index (*PCI*), an integrative variable of stomatal density and the length of guard cells, which can be used as a proxy for the theoretical maximal stomatal water vapour conductance [56,57]:

$$PCI = L_A^2 \times SD \times 10^{-4} \tag{5}$$

L_A—length of guard cells, SD—stomatal density.

Specific leaf area (*SLA*), a parameter that corresponds to the thickness and density of leaf lamina [58,59]:

$$SLA = \frac{S_{leaf}}{m_{leaf}} \tag{6}$$

S_{leaf}—leaf size and m_{leaf}—dry weight of leaves.

2.3. Quantification of Phenotypic Plasticity

The plasticity index based on maximum and minimum means (PI_v) was calculated for each trait and provenance, respectively [37].

$$PI_v = \frac{\overline{x}_{max} - \overline{x}_{min}}{\overline{x}_{max}} \tag{7}$$

$\overline{x}_{max}$—maximum mean (mean of the group showing the maximal value relative to that of other groups); $\overline{x}_{min}$—minimum mean (mean of the group showing the minimal value relative to that of other groups). In our case, the groups refer to identical provenances from two different provenance plots.

2.4. Statistical Analysis

Statistical analysis was performed using R statistical software (Version 4.0.3, R Core Team, Vienna, Austria). Original climate characteristics of provenances were analysed by principal component analysis to visualize likeness or disparity between the original sites of the provenances. The normal distribution of the obtained data was first tested by the Shapiro–Wilk test. A two-way analysis of variance was used where provenance and plots were set as factors with fixed effects. Moreover, differences between provenances were tested separately for each plot by Fisher's LSD post hoc test. We used ecodistance to capture shifts between the climate of origin and the climate of the provenance plot. The ecodistance was defined as the difference between the investigated ecologically relevant variables at the test site and at the population origin [22,60]. Furthermore, mixed models in the R "nlme" package [61] were used to explore relationships between individual stomatal and leaf morphological traits (dependent variables) and various explanatory variables, such as *EQ*, *FAI*, latitude and longitude. To account for between-plot variability, we included plot as a random effect variable, and the between-plot variance in the intercept estimation was quantified. We were not able to include a block design in the mixed models, as there were missing data due to spatial mortality within the plots. We used maximum likelihood to estimate the parameters of the model. To estimate the importance of individual explanatory variables, we calculated the normalized model likelihoods (Akaike weights). First, we fitted all possible model variants, including the null model (including only the intercept) and full model (including all explanatory variables). Then, Akaike weights were calculated for each model based on the corrected Akaike information criterion using the "Weights" function in the "MuMIn" R package [62]. The models with the highest weights were further selected and interpreted. In addition to that, we calculated marginal R^2 (R^2_m) and conditional R^2 (R^2_c) for better comparison of fixed and random factors in model [63]. Statistical significance of differences between the plasticity of the traits was assessed by analysis of variance and Tukey's post-hoc test. The relationships between original climate, phenotypic plasticity and tree mortality were tested by linear regression.

3. Results

3.1. Stomatal and Leaf Morphological Traits

The effects of the tested factors, provenance, plot and provenance-by-plot interactions were statistically significant for all traits except in the case of specific leaf area (*SLA*) (Table 2, Table A1 in Appendix A). Provenances growing in a warmer and drier site in the Czech Republic showed lower guard cell length (L_A, Figure 2A), stomatal density (*SD*, Figure 2B), potential conductance index (*PCI*, Figure 2C), leaf dry weight (m_{leaf}, Figure 2D) and leaf size (S_{leaf}, Figure 2E) values than the provenances in the colder and more humid Slovak plot. Provenances in the Czech provenance plot showed an average 57% reduction in the *PCI* for leaves that were 12% smaller than those in the Slovak provenance plot. Based on the average m_{leaf}, the provenances in the drier Czech plot accumulated 39% less biomass per leaf than the provenances in the more humid Slovak plot. Values of *SLA* were higher on average for the drier provenance plot, but changes were inconsistent and insignificant between provenances (Figure 2F).

Table 2. Results of two-way ANOVAs for each stomatal and leaf morphological trait.

Factor	Df	Trait	L_A	SD	PCI	m_{leaf}	S_{leaf}	SLA
Provenance	1	F	57.95 ***	17.65 ***	11.59 ***	3.79 **	10.83 ***	0.69
		p						0.64
Plot	8	F	1771.5 ***	457.4 ***	1221.25 ***	31.35 ***	21.58 ***	9.48
		p						**
Provenance × Plot	8	F	18.09 ***	4.642 ***	7.66 ***	2.72 *	3.92 ***	1.64
		p						0.13

L_A—length of guard cells, *SD*—stomatal density, *PCI*—potential conductance index, S_{leaf}—leaf size, m_{leaf}—dry weight per leaf, *SLA*—specific leaf area, significance levels *** <0.001 ** <0.01 * <0.05.

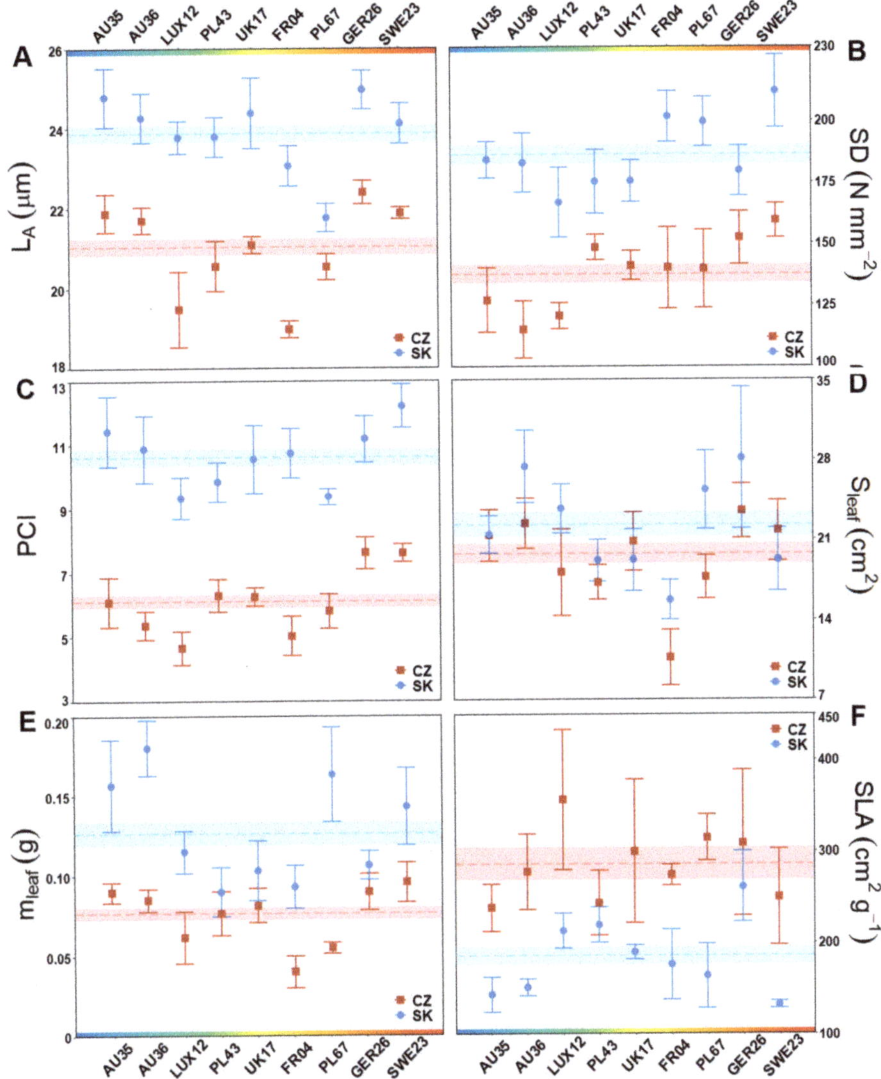

Figure 2. Mean values and standard errors of stomatal and leaf morphological traits per provenance: length of guard cell (L_A, (**A**)), stomatal density (*SD*, (**B**)), potential conductance index (*PCI*, (**C**)), leaf size (S_{leaf}, (**D**)), leaf dry weight (m_{leaf}, (**E**)) and specific leaf area (*SLA*, (**F**)). The red colour represents provenances growing in the Czech provenance plot, and the blue colour represents provenances growing in the Slovak provenance plot. The dashed horizontal line represents the average per plot with the surrounding standard error interval band. Provenances are arranged based on the Ellenberg quotient of the original site (blue to red *x* axis band).

3.2. Impact of Climate Ecodistance on Stomatal and Leaf Morphological Traits

The multifactorial approach based on mixed models showed that the models with singular explanatory variables performed better than those with multiple factors; thus, a further analysis employed simple linear regression models (Tables A2 and A3). The degree of aridity and temperature differences between provenances' origins and new plots (ecodistance) had a significant effect on provenances' stomatal development (Figure 3). The ecodistance of the Ellenberg quotient (EQ_{ED}) and forest aridity index (FAI_{ED}) had a significant negative influence on SD (Figure 3A,B) and the PCI (Figure 3C,D). Provenances transferred to a drier environment relative to their original site showed a proportionally lower density of stomata with lower potential conductance. The ecodistance of average temperature (T_{ED}) and average temperature during the vegetation season (T_{59ED}) also had a significant negative influence on SD, and provenances transferred to a climate warmer than that of their original site showed a proportional decrease in SD (Figure 3E,F). Other climate ecodistance indices showed no significant correlations with stomatal and leaf morphological traits.

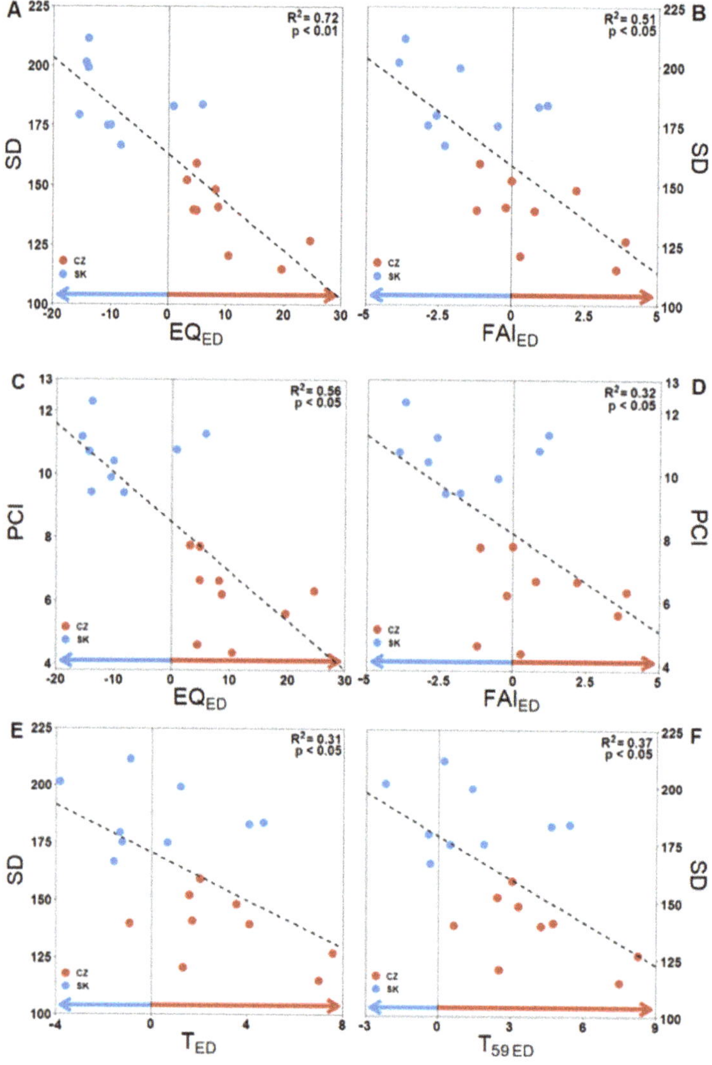

Figure 3. Linear regression results showing the relationship between ecodistance (EQ_{ED}, FAI_{ED}, T_{ED}, $T_{59\,ED}$) and stomatal morphological traits, stomatal density (SD, (**A,B,E,F**)) and the potential conductance index (PCI, (**C,D**)). Positive values on the horizontal axis represent transfer to a drier/hotter environment (red arrow), and negative values represent transfer to a more humid/colder environment (blue arrow).

3.3. Phenotypic Plasticity of Stomatal and Leaf Morphological Traits

To quantify the acclimation response, we calculated the phenotypic plasticity index for each trait and provenance. Plasticity indices of the potential conductance index (PCI_{PI}), leaf dry weight ($m_{leaf\,PI}$) and specific leaf area (SLA_{PI}) showed the highest values of plasticity among the provenances. In contrast, the plasticity index of the length of guard cells ($L_{A\,PI}$) was the lowest among the provenances (Figure 4). The climate of the original provenance locations affected the plasticity response of acclimation after transfer to a new environment. Provenances from locations with more heterogeneous environmental temperatures showed lower PCI_{PI} and $L_{A\,PI}$ plasticity. We found a positive relationship between the isothermality (Iso_T) of the original location and provenances PCI_{PI} and $L_{A\,PI}$ (Figure 5A,C). Moreover, provenances from environments with more heterogeneous precipitation distributions showed lower plasticity of L_A. The seasonality of precipitation ($Seas_{Prec}$) at the original locations of the provenances negatively influenced the $L_{A\,PI}$ of the provenances (Figure 5D). Furthermore, the original latitude (Lat) affected the PCI_{PI}, as provenances from lower latitudes showed a higher PCI_{PI} (Figure 5B). We found a significant negative relationship between provenance mortality and the plasticity of stomatal and leaf morphological traits (PCI_{PI} and SLA_{PI}) in the drier and hotter Czech provenance plot (Figure 6) but no significant relationship between mortality and plasticity in the more humid Slovak provenance plot.

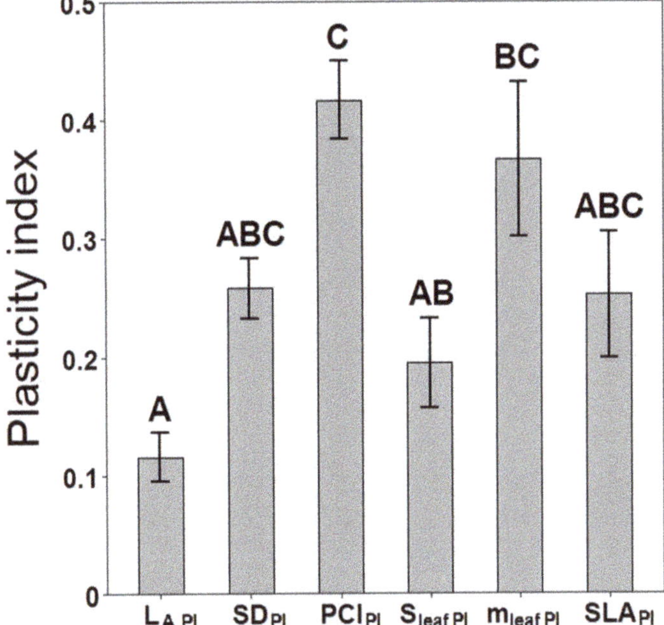

Figure 4. Phenotypic plasticity indices of stomatal density (SD_{PI}), guard cell length ($L_{A\,PI}$), the potential conductance index (PCI_{PI}), leaf size ($S_{leaf\,PI}$), leaf dry weight ($m_{leaf\,PI}$) and specific leaf area (SLA_{PI}) for each provenance. Plasticity indices obtain values of 0 to 1, where 0 denotes no plasticity and 1 denotes theoretical maximal plasticity. Confidence intervals represent standard error and the capital letters correspond to results of post-hoc analysis.

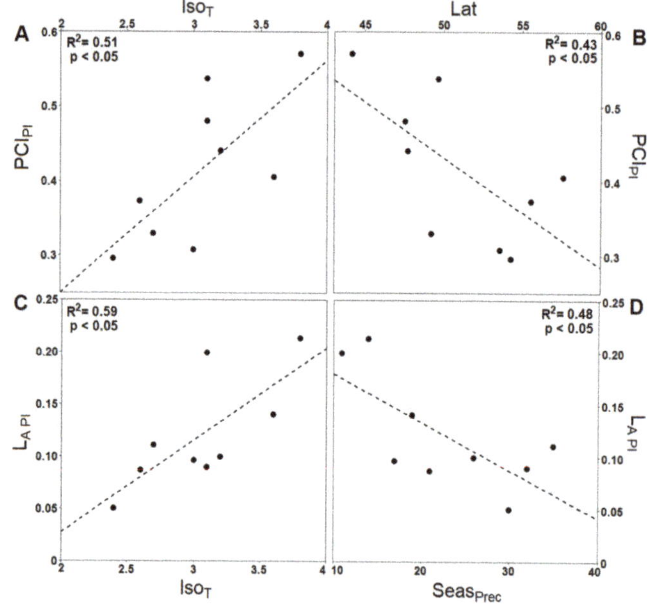

Figure 5. Linear regression results showing the relationship between isothermality (Iso_T, (**A**,**C**)), latitude (*Lat*, (**B**)) and precipitation seasonality ($Seas_{Prec}$, (**D**)) of the provenance's original site and plasticity indices ($L_{A\ PI}$, PCI_{PI}).

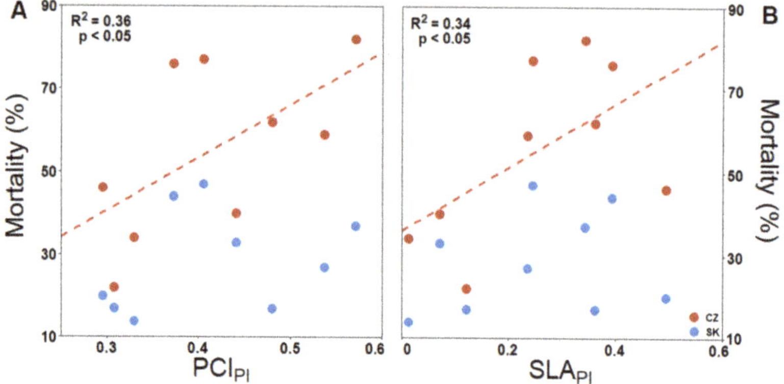

Figure 6. Visualization of the relationship between plasticity indices (PCI_{PI}, (**A**); SLA_{PI}, (**B**)) and the observed mortality of provenances in the Czech (red) and Slovak (blue) provenance plot. The linear regression was significant only for drier Czech provenance plot.

4. Discussion

4.1. Functional Aspects of the Adaptive Response

Our study demonstrates that European beech provenances adjusted their stomatal and leaf morphological traits in response to being transferred to a new environment. Provenances exhibited significantly lower SD, L_A, PCI, S_{leaf} and m_{leaf} levels in the drier Czech provenance plot than in the more humid Slovak provenance plot. There were no significant differences in SLA when we considered provenance–plot interactions. Lack of significant differences regarding SLA might be caused by low sample size and high variability of the trait. Nevertheless, we observed that provenances in more xeric plot showed higher overall *SLA*. This is in opposition to general consensus [13,36,64,65], but has already been observed in some studies [66,67]. The adjustment of stomatal and leaf development can be seen as an adaptive response to either suboptimal climatic conditions of the Czech plot or to favourable climatic conditions of the Slovak plot. It has been reported that plants might improve their drought tolerance and water use efficiency (*WUE*) by reducing *SD* [31,32,68] and L_A [28,69]. Both herbaceous plants and trees react to episodic drought and long-term xericity of the environment by decreasing *SD* [25,70] and developing smaller stomata with lower L_A [26,56,71–73]. Combined stomatal morphology (L_A) and the distribution of stomata on leaves (*SD*), represented as the *PCI*, might be seen as a proxy for structural constraints of maximal stomatal conductance. A reduction in the *PCI* under xeric conditions should then ultimately reduce stomatal conductance [74] and water loss, which can lead to improved *WUE* [73,75]. Acclimation through the development of smaller leaves (S_{leaf}) under xeric conditions leads to less water loss through transpiration [76] and higher *WUE* [77,78]. It should be mentioned that *WUE* is also influenced by photosynthetic capacity and not just stomata related traits [13]. Plants exposed to water deficit show a reduction in m_{leaf} [64,79]. The combination of lower S_{leaf} with m_{leaf} in drier site might be explained by trees' strategy to invest more in root biomass with the cost of lower leaf biomass and leaf size [80]. The above-mentioned traits, therefore, represent plants' adaptive mechanisms in mitigating drought stress [81]. As these stomatal and leaf morphological traits have a significant impact on plant performance under water stress, their adjustment is vital for plants to successfully acclimate under changing conditions due to either anthropogenic transport to new environments or accelerating global climate change.

On the other hand, an increase in *SD* and L_A can enhance photosynthetic capacity [82,83], which could enhance tree performance under optimal climatic conditions where the strongest selective pressure is competition [84]. Leaves with higher *SD* and larger stomata (L_A) show an increase in maximal stomatal conductance [85–87], which leads to higher biomass accumulation and growth [88]. A higher *PCI* increases the maximal limit of stomatal conductance, which might improve photosynthetic capacity [89]. Moreover, the *PCI* has also been found to be related to leaf hydraulic conductance [57], which has been correlated with photosynthesis rates across plant species [89,90]. Higher S_{leaf} values under favourable conditions lead to higher photosynthesis rates [91,92], which positively affect leaf biomass production with higher m_{leaf} values [93]. According to our results, provenances in the drier and hotter Czech plot might acclimatise to their environment by undertaking a water conservation strategy with decreasing *SD*, L_A, *PCI*, S_{leaf} and m_{leaf} values. Conversely, provenances growing on the more humid Slovak plot could utilize the development of larger leaves (S_{leaf}, m_{leaf}) with higher *SD*, L_A and *PCI* values to maximize photosynthetic activity and growth in a competitive environment. Despite the clear theoretical basis for why the provenances showed significantly different values of the tested traits, we did not find any significant relationship between the tested traits and mortality. To address our first hypothesis (i), we found the morphological response as expected, but there is no evidence that the alternation of stomatal and leaf morphological traits had a significant positive effect on provenance performance. To capture the drought resistance profile of provenances for practical application, we suggest analysing additional physiological and functional traits, such as *WUE*, cuticular conductance, xylem embolism resistance and the turgor loss point [10,13,94]. Stomatal and leaf morphological traits alone are not satisfactory to define which provenances would be favourable for hotter and more xeric conditions in the near future.

4.2. Climate Ecodistance as an Effective Tool for Provenance Research

We found that the *SD* and *PCI* of provenances depend on the aridity ecodistance (EQ_{ED} and FAI_{ED}) and temperature ecodistance (T_{ED} and T_{59ED}). The ecodistance represents the climatic shift between the original provenance site and the provenance plot to which it is transferred. The connection between both the climate of origin and the current climate and stomatal phenotype might suggest strong coordination between genetic and environmental impacts on stomatal development. A previous provenance study showed strong significant relationships between temperature and ecodistance and the phenology, morphology and dendrometric traits of European beech provenances [22]. Aridity ecodistance (*EQ*) has been found to be a significant explanatory variable for the vitality [95] and growth [96] of European beech provenances. A study of four temperate tree species (*Fagus sylvatica*, *Picea abies* (L.) Karst., *Pinus sylvestris* L. and *Quercus petraea* (Matt.) Liebl.) has also shown a significant relationship between aridity ecodistance (the annual aridity index) and the growth of provenances after transfer to a new environment [97]. Our second hypothesis (ii) is confirmed by a significant relationship between *SD*, the *PCI* and ecodistance, which incorporates the climate of origin as well as the current climate of new plots. The ecodistance has not attracted much popularity since its first formulation [22], but it seems that this simple mathematical formulation of climate transfer could capture the physiological, morphological and growth reactions of tree provenances to their new environments. More studies have focused on additional tree species, and other aspects of tree phenotyping are needed to test whether ecodistance is robust enough to be useful and reliable for forestry applications. The ecodistance could then be used not only to explain the effect of climatic shifts caused by the spatial transfer of provenances but also to predict provenance reactions to temporal changes caused by accelerating anthropogenic global climate change.

4.3. Phenotypic Plasticity

Populations with higher phenotypic plasticity can adapt to higher environmental variability and thus can minimize the risk of mortality [65,98,99]. Our results suggest that the *PCI*, m_{leaf} and *SLA* are the most plastic, while L_A is the least plastic trait among provenances. The low plasticity of L_A might suggest higher genetic control of stomatal size relative to the other tested traits. Similar results of low plasticity for L_A and high plasticity for the *PCI* and leaf morphology were also observed in beech provenance studies [71,95]. Furthermore, we found a significant relationship between climate heterogeneity of the original site (Iso_T and $Seas_{Prec}$) and the plasticity of L_A and *PCI* ($L_{A\,PI}$ and PCI_{PI}). We expected populations that had evolved under more heterogeneous environments to favour higher phenotypic plasticity to quickly adjust their phenotype if needed [11,37,99,100]. In contrast to what we expected, the provenances from the most heterogeneous environment showed the lowest $L_{A\,PI}$ and PCI_{PI} values, so hypothesis (iii) cannot be confirmed. We also found a negative significant relationship between latitude and PCI_{PI}, which might be attributable to a reduction in genetic diversity from lower to higher latitudes after the recolonization of habitats after the last glacial maximum [101–103]. A harsher northern environment or higher competition at the distribution edge may create more selection pressure than elsewhere in the distribution range [44]. Therefore, populations that evolved under strong selection pressure might favour more efficient strong genetic control over high phenotypic plasticity within the population [104]. This could lead to trait canalization which might be translated to lower phenotypic plasticity [105–107].

We found a significant positive relationship between PCI_{PI}, SLA_{PI} and mortality values, but only in the drier and hotter Czech provenance plots. However, in this study with phenotyping data from only the remaining trees, we cannot conclude that the high plasticity of these two traits causes high provenance mortality as the plasticity was calculated from surviving individuals and thus might be biased. Higher mortality, which created more open canopies and less competition among the remaining trees, might have caused the higher values of phenotypic plasticity, as trees reacted to newly available canopy space. High phenotypic plasticity is generally seen as a favourable property of plants, trees or populations under global climate change [37,108,109]. We cannot confirm our hypothesis that (iv) there is a negative relationship between phenotypic plasticity and tree mortality. Despite this, we find it important to discuss phenotypic plasticity. Both recent and earlier studies have shown that the high phenotypic plasticity of plants should not be universally seen as a positive attribute [43,45,110,111]. The plasticity cost might not be pronounced under normal conditions, but when plants are exposed to a highly stressful environment, the plasticity cost might outweigh the fitness gain [112,113]. The results of several studies suggest reduced performance with increasing phenotypic plasticity under stress [38,44]. Thus, phenotypic plasticity, measured by common metrics [37], should not be automatically interpreted as beneficial for plants under global climate change. The high phenotypic plasticity of populations exposed to severe environmental stress might be associated with increased mortality and reduced fitness.

5. Conclusions

European beech provenances have shown a high degree of both adaptation and acclimation after transfer to a new environment. The observed differences in stomatal morphological traits were linked to the long-term aridity and air temperature of both the original site and the current provenance plot. The heterogeneity of the original site's climate and latitude affected the phenotypic plasticity of stomatal traits. Higher phenotypic plasticity was associated with higher mortality under suboptimal conditions but not under favourable hydric conditions. Additional functional and physiological traits should be considered to evaluate the resistance or performance of European beech provenances, as we have not found any direct link between mortality and the tested stomatal and leaf morphological traits. Ecodistance can be considered as an easy to use and robust tool for analysing adaptive responses of tree provenances under global climate change. Studies of phenotypic plasticity should not interpret the positive effects of high plasticity without taking into consideration the performance, fitness or vitality of plants.

Author Contributions: D.K., J.F. and A.K. conceived the experimental design and coordinated the experiment; P.P., A.P., A.K., J.F. and D.K. sampled the plant material; A.P. and P.P. processed the samples and performed the measurements; M.B., P.P. and P.F. performed the statistical analyses; P.P. and A.P. wrote the first version of the manuscript, which was discussed and approved by all of the co-authors. All authors have read and agreed to the published version of the manuscript.

Funding: The study was supported by research grants from the Slovak Research and Development Agency APVV-18-0390, by the Slovak Grant Agency for Science no. VEGA 1/0535/20 and by the Ministry of Agriculture of the Czech Republic, institutional support MZE-RO0118. Climatic data provided by SHMÚ and ČHMÚ are greatly appreciated.

Conflicts of Interest: The authors declare no conflict of interest.

Appendix A

Table A1. Results of Fisher LSD post-hoc test.

SK	L_A	SD	PCI	S_{leaf}	m_{leaf}	SLA
FR04	BC	AC	AB	B	A	AB
LUX12	AB	B	A	AB	ABC	BC
UK17	AB	AB	AB	AB	AB	ABC
SWE23	AB	C	B	AB	ABCD	A
GER26	A	ABC	AB	A	ABC	C
AU35	A	ABC	AB	AB	BCD	A
AU36	AB	ABC	AB	A	D	AB
PL43	AB	AB	A	AB	A	BC
PL67	C	AC	A	A	CD	AB

CZ	L_A	SD	PCI	S_{leaf}	m_{leaf}	SLA
FR04	C	ABCD	AB	B	D	A
LUX12	C	AB	B	AB	ACD	A
UK17	AB	ABCD	ACD	A	ABC	A
SWE23	AB	D	CD	A	B	A
GER26	B	CD	D	A	AB	A
AU35	AB	ABC	AC	A	AB	A
AU36	AB	A	AB	A	ABC	A
PL43	AC	BCD	ACD	AB	ABCD	A
PL67	AC	ABCD	AB	AB	CD	A

SD—stomatal density, L_A—length of guard cells, PCI—potential conductance index, SLA—specific leaf area, S_{leaf}—leaf size, m_{leaf}—dry weight per leaf.

Table A2. AICc based variable weights of mixed models. Green to red spectrum represents the explanatory power of individual parameters and their combinations. Full model consists of all three parameters: aridity (expressed as the *EQ* and *FAI*), longitude and latitude.

Trait	EQ_{ED}	$Long_{ED}$	Lat_{ED}	$EQ_{ED} + Long_{ED}$	$EQ_{ED} + Lat_{ED}$	$Long_{ED} + Lat_{ED}$	Full Model
SD	0.704	0.014	0.007	0.125	0.131	0.002	0.016
L_A	0.222	0.31	0.168	0.202	0.031	0.047	0.021
PCI	0.211	0.461	0.151	0.065	0.032	0.073	0.007
SLA	0.312	0.239	0.292	0.044	0.06	0.046	0.006
S_{leaf}	0.16	0.472	0.152	0.069	0.031	0.102	0.014
	FAI_{ED}	$Long_{ED}$	Lat_{ED}	$FAI_{ED} + Long_{ED}$	$FAI_{ED} + Lat_{ED}$	$Long_{ED} + Lat_{ED}$	Full model
SD	0.263	0.056	0.029	0.046	0.532	0.009	0.065
L_A	0.286	0.284	0.154	0.163	0.049	0.043	0.021
PCI	0.177	0.482	0.158	0.068	0.031	0.076	0.008
SLA	0.316	0.241	0.296	0.047	0.048	0.047	0.005
S_{leaf}	0.149	0.345	0.111	0.273	0.021	0.075	0.027

SD—stomatal density, L_A—length of guard cells, *PCI*—potential conductance index, *SLA*—specific leaf area, S_{leaf}—leaf size, EQ_{ED}—Ellenberg quotient ecodistance, FAI_{ED}—Forest Aridity Index ecodistance, $Long_{ED}$—longitude ecodistance, Lat_{ED}—latitude ecodistance.

Table A3. Values of the marginal and conditional R^2 of mixed models presented in Table A2.

	SD		L_A		*PCI*		*SLA*		S_{leaf}	
	R^2_m	R^2_c	R^2_m	R^2_c	R^2_m	R^2_c	R^2_m	R^2_c	R^2_m	R^2_c
EQ_{ED}	0.38	0.76	0.03	0.87	0.01	0.89	0.04	0.86	0.02	0.38
$Long_{ED}$	0.01	0.84	0.02	0.78	0.01	0.91	0.01	0.75	0.12	0.18
Lat_{ED}	0.00	0.84	0.00	0.78	0.00	0.91	0.01	0.77	0.03	0.22
$EQ_{ED} + Long_{ED}$	0.38	0.75	0.07	0.92	0.01	0.91	0.04	0.86	0.19	0.75
$EQ_{ED} + Lat_{ED}$	0.37	0.77	0.03	0.86	0.01	0.88	0.04	0.86	0.05	0.40
$Long_{ED} + Lat_{ED}$	0.02	0.84	0.02	0.78	0.01	0.91	0.02	0.77	0.15	0.29
Full model	0.37	0.76	0.07	0.91	0.01	0.91	0.04	0.86	0.21	0.77

	SD		L_A		*PCI*		*SLA*		S_{leaf}	
	R^2_m	R^2_c	R^2_m	R^2_c	R^2_m	R^2_c	R^2_m	R^2_c	R^2_m	R^2_c
FAI_{ED}	0.06	0.82	0.02	0.83	0.00	0.90	0.02	0.80	0.09	0.43
$Long_{ED}$	0.01	0.84	0.02	0.78	0.01	0.91	0.01	0.75	0.12	0.18
Lat_{ED}	0.00	0.84	0.00	0.78	0.00	0.91	0.01	0.77	0.03	0.22
$FAI_{ED} + Long_{ED}$	0.06	0.82	0.04	0.85	0.01	0.91	0.02	0.79	0.23	0.58
$FAI_{ED} + Lat_{ED}$	0.11	0.86	0.02	0.83	0.00	0.90	0.02	0.79	0.09	0.40
$Long_{ED} + Lat_{ED}$	0.02	0.84	0.02	0.78	0.01	0.91	0.02	0.77	0.15	0.29
Full model	0.11	0.86	0.05	0.85	0.01	0.90	0.02	0.78	0.23	0.56

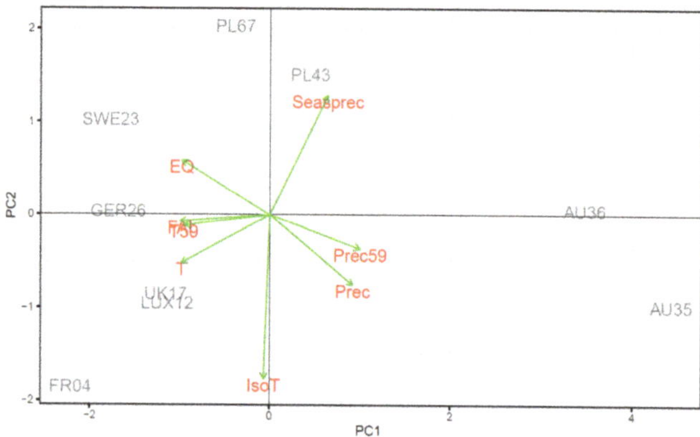

Figure A1. The principal component analysis biplot of provenance's original climate.

References

1. Leuschner, C.; Backes, K.; Hertel, D.; Schipka, F.; Schmitt, U.; Terborg, O.; Runge, M. Drought responses at leaf, stem and fine root levels of competitive Fagus sylvatica L. and Quercus petraea (Matt.) Liebl. trees in dry and wet years. *For. Ecol. Manag.* **2001**, *149*, 33–46. [CrossRef]
2. Bréda, N.; Huc, R.; Granier, A.; Dreyer, E. Temperate forest trees and stands under severe drought: A review of ecophysiological responses, adaptation processes and long-term consequences. *Ann. For. Sci.* **2006**, *63*, 625–644. [CrossRef]
3. Gessler, A.; Keitel, C.; Kreuzwieser, J.; Matyssek, R.; Seiler, W.; Rennenberg, H. Potential risks for European beech (Fagus sylvatica L.) in a changing climate. *Trees* **2006**, *21*, 1–11. [CrossRef]
4. Kramer, K.; Degen, B.; Buschbom, J.; Hickler, T.; Thuiller, W.; Sykes, M.T.; De Winter, W. Modelling exploration of the future of European beech (Fagus sylvatica L.) under climate change—Range, abundance, genetic diversity and adaptive response. *For. Ecol. Manag.* **2010**, *259*, 2213–2222. [CrossRef]
5. Hacket-Pain, A.J.; Cavin, L.; Friend, A.D.; Jump, A.S. Consistent limitation of growth by high temperature and low precipitation from range core to southern edge of European beech indicates widespread vulnerability to changing climate. *Eur. J. For. Res.* **2016**, *135*, 897–909. [CrossRef]
6. Piovesan, G.; Biondi, F.; Di Filippo, A.; Alessandrini, A.; Maugeri, M. Drought-driven growth reduction in old beech (Fagus sylvatica L.) forests of the central Apennines, Italy. *Glob. Chang. Biol.* **2008**, *14*, 1265–1281. [CrossRef]
7. Charru, M.; Seynave, I.; Morneau, F.; Bontemps, J.-D. Recent changes in forest productivity: An analysis of national forest inventory data for common beech (Fagus sylvatica L.) in north-eastern France. *For. Ecol. Manag.* **2010**, *260*, 864–874. [CrossRef]
8. Rita, A.; Gentilesca, T.; Ripullone, F.; Todaro, L.; Borghetti, M. Differential climate–growth relationships in Abies alba Mill. and Fagus sylvatica L. in Mediterranean mountain forests. *Dendrochronologia* **2014**, *32*, 220–229. [CrossRef]
9. Rose, L.; Leuschner, C.; Köckemann, B.; Buschmann, H. Are marginal beech (Fagus sylvatica L.) provenances a source for drought tolerant ecotypes? *Eur. J. For. Res.* **2009**, *128*, 335–343. [CrossRef]
10. Schuldt, B.; Knutzen, F.; Delzon, S.; Jansen, S.; Müller-Haubold, H.; Burlett, R.; Clough, Y.; Leuschner, C. How adaptable is the hydraulic system of European beech in the face of climate change-related precipitation reduction? *New Phytol.* **2015**, *210*, 443–458. [CrossRef]
11. Aranda, I.; Bahamonde, H.A.; Sánchez-Gómez, D. Intra-population variability in the drought response of a beech (Fagus sylvatica L.) population in the southwest of Europe. *Tree Physiol.* **2017**, *37*, 938–949. [CrossRef] [PubMed]
12. Robson, T.M.; BeechCOSTe52 Database Consortium; Garzón, M.B. Phenotypic trait variation measured on European genetic trials of *Fagus sylvatica* L. *Sci. Data* **2018**, *5*, 180149. [CrossRef] [PubMed]
13. Sánchez-Gómez, D.; Robson, T.M.; Gascó, A.; Gil-Pelegrín, E.; Aranda, I. Differences in the leaf functional traits of six beech (Fagus sylvatica L.) populations are reflected in their response to water limitation. *Environ. Exp. Bot.* **2013**, *87*, 110–119. [CrossRef]
14. García-Plazaola, J.I.; Becerril, J. Effects of drought on photoprotective mechanisms in European beech (Fagus sylvatica L.) seedlings from different provenances. *Trees* **2000**, *14*, 485–490. [CrossRef]
15. Peuke, A.D.; Schraml, C.; Hartung, W.; Rennenberg, H. Identification of drought-sensitive beech ecotypes by physiological parameters. *New Phytol.* **2002**, *154*, 373–387. [CrossRef]
16. Bolte, A.; Czajkowski, T.; Cocozza, C.; Tognetti, R.; De Miguel, M.; Pšidová, E.; Ditmarová, Ľ.; Dinca, L.; Delzon, S.; Cochard, H.; et al. Desiccation and Mortality Dynamics in Seedlings of Different European Beech (Fagus sylvatica L.) Populations under Extreme Drought Conditions. *Front. Plant Sci.* **2016**, *7*, 751. [CrossRef]
17. Bresson, C.C.; Vitasse, Y.; Kremer, A.; Delzon, S. To what extent is altitudinal variation of functional traits driven by genetic adaptation in European oak and beech? *Tree Physiol.* **2011**, *31*, 1164–1174. [CrossRef]
18. Dounavi, A.; Netzer, F.; Ćelepirović, N.; Ivanković, M.; Burger, J.; Figueroa, A.; Schön, S.; Simon, J.; Cremer, E.; Fussi, B.; et al. Genetic and physiological differences of European beech provenances (F. sylvatica L.) exposed to drought stress. *For. Ecol. Manag.* **2016**, *361*, 226–236. [CrossRef]
19. Thiel, D.; Kreyling, J.; Backhaus, S.; Beierkuhnlein, C.; Buhk, C.; Egen, K.; Huber, G.; Konnert, M.; Nagy, L.; Jentsch, A. Different reactions of central and marginal provenances of Fagus sylvatica to experimental drought. *Eur. J. For. Res.* **2014**, *133*, 247–260. [CrossRef]

20. Aranda, I.; Cano, F.J.; Gascó, A.; Cochard, H.; Nardini, A.; Mancha, J.A.; López, R.; Sánchez-Gómez, D. Variation in photosynthetic performance and hydraulic architecture across European beech (Fagus sylvatica L.) populations supports the case for local adaptation to water stress. *Tree Physiol.* **2015**, *35*, 34–46. [CrossRef]

21. De Villemereuil, P.; Gaggiotti, O.E.; Mouterde, M.; Till-Bottraud, I. Common garden experiments in the genomic era: New perspectives and opportunities. *Heredity* **2016**, *116*, 249–254. [CrossRef] [PubMed]

22. Matyas, C. Climatic adaptation of trees: Rediscovering provenance tests. *Euphytica* **1996**, *92*, 45–54. [CrossRef]

23. Konôpková, A.; Kurjak, D.; Kmeť, J.; Klumpp, R.; Longauer, R.; Ditmarová, Ľ.; Gömöry, D. Differences in photochemistry and response to heat stress between silver fir (Abies alba Mill.) provenances. *Trees* **2017**, *32*, 73–86. [CrossRef]

24. Kurjak, D.; Konôpková, A.; Kmeť, J.; Macková, M.; Frýdl, J.; Zivcak, M.; Palmroth, S.; Ditmarová, Ľ.; Gömöry, D. Variation in the performance and thermostability of photosystem II in European beech (Fagus sylvatica L.) provenances is influenced more by acclimation than by adaptation. *Eur. J. For. Res.* **2018**, *138*, 79–92. [CrossRef]

25. Lavilla, M.; Seral, A.; Murciano, A.; Molino, S.; De-La-Fuente-Redondo, P.; Galan, J.M.G.Y. Stomatal traits in Iberian populations of Osmunda regalis (Osmundaceae, Polypodiopsida) and its relationship with bioclimatic variables. *Acta Bot. Malacit.* **2018**, *42*, 5–13. [CrossRef]

26. Durand, M.; Brendel, O.; Buré, C.; Le Thiec, D. Changes in irradiance and vapour pressure deficit under drought induce distinct stomatal dynamics between glasshouse and field-grown poplars. *New Phytol.* **2020**, *227*, 392–406. [CrossRef]

27. Gindel, I. Stomatal Number and Size as Related to Soil Moisture in Tree Xerophytes in Israel. *Ecology* **1969**, *50*, 263–267. [CrossRef]

28. Drake, P.L.; Froend, R.H.; Franks, P.J. Smaller, faster stomata: Scaling of stomatal size, rate of response, and stomatal conductance. *J. Exp. Bot.* **2013**, *64*, 495–505. [CrossRef]

29. McCree, K.J.; Davis, S.D. Effect of Water Stress and Temperature on Leaf Size and on Size and Number of Epidermal Cells in Grain Sorghum 1. *Crop. Sci.* **1974**, *14*, 751–755. [CrossRef]

30. Cutler, J.M.; Rains, D.W.; Loomis, R.S. The Importance of Cell Size in the Water Relations of Plants. *Physiol. Plant.* **1977**, *40*, 255–260. [CrossRef]

31. Hepworth, C.; Doheny-Adams, T.; Hunt, L.; Cameron, D.D.; Gray, J.E. Manipulating stomatal density enhances drought tolerance without deleterious effect on nutrient uptake. *New Phytol.* **2015**, *208*, 336–341. [CrossRef] [PubMed]

32. Caine, R.S.; Yin, X.; Sloan, J.; Harrison, E.L.; Mohammed, U.; Fulton, T.; Biswal, A.K.; Dionora, J.; Chater, C.C.; Coe, R.A.; et al. Rice with reduced stomatal density conserves water and has improved drought tolerance under future climate conditions. *New Phytol.* **2018**, *221*, 371–384. [CrossRef] [PubMed]

33. Poorter, H.; Niinemets, Ü.; Poorter, L.; Wright, I.J.; Villar, R. Causes and consequences of variation in leaf mass per area (LMA): A meta-analysis. *New Phytol.* **2009**, *182*, 565–588. [CrossRef] [PubMed]

34. Wright, I.J.; Reich, P.B.; Westoby, M. Strategy shifts in leaf physiology, structure and nutrient content between species of high- and low-rainfall and high- and low-nutrient habitats. *Funct. Ecol.* **2001**, *15*, 423–434. [CrossRef]

35. Wellstein, C.; Chelli, S.; Campetella, G.; Bartha, S.; Galiè, M.; Spada, F.; Canullo, R. Intraspecific phenotypic variability of plant functional traits in contrasting mountain grasSLAnds habitats. *Biodivers. Conserv.* **2013**, *22*, 2353–2374. [CrossRef]

36. Wellstein, C.; Poschlod, P.; Gohlke, A.; Chelli, S.; Campetella, G.; Rosbakh, S.; Canullo, R.; Kreyling, J.; Jentsch, A.; Beierkuhnlein, C. Effects of extreme drought on specific leaf area of grasSLAnd species: A meta-analysis of experimental studies in temperate and sub-Mediterranean systems. *Glob. Chang. Biol.* **2017**, *23*, 2473–2481. [CrossRef]

37. Valladares, F.; Sanchez-Gomez, D.; Zavala, M.A. Quantitative estimation of phenotypic plasticity: Bridging the gap between the evolutionary concept and its ecological applications. *J. Ecol.* **2006**, *94*, 1103–1116. [CrossRef]

38. Merilä, J.; Hendry, A.P. Climate change, adaptation, and phenotypic plasticity: The problem and the evidence. *Evol. Appl.* **2014**, *7*, 1–14. [CrossRef]

39. Stamp, M.A.; Hadfield, J.D. The relative importance of plasticity versus genetic differentiation in explaining between population differences; a meta-analysis. *Ecol. Lett.* **2020**, *23*, 1432–1441. [CrossRef]

40. Palacio-López, K.; Beckage, B.; Scheiner, S.M.; Molofsky, J. The ubiquity of phenotypic plasticity in plants: A synthesis. *Ecol. Evol.* **2015**, *5*, 3389–3400. [CrossRef]

41. Blackman, C.J.; Aspinwall, M.J.; Tissue, D.T.; Rymer, P.D. Genetic adaptation and phenotypic plasticity contribute to greater leaf hydraulic tolerance in response to drought in warmer climates. *Tree Physiol.* **2017**, *37*, 583–592. [CrossRef] [PubMed]

42. Kramer, K.; Ducousso, A.; Gömöry, D.; Hansen, J.K.; Ionita, L.; Liesebach, M.; Lorenţ, A.; Schueler, S.; Sułkowska, M.; De Vries, S.; et al. Chilling and forcing requirements for foliage bud burst of European beech (Fagus sylvatica L.) differ between provenances and are phenotypically plastic. *Agric. For. Meteorol.* **2017**, *2017*, 172–181. [CrossRef]

43. DeWitt, T.J.; Sih, A.; Wilson, D.S. Costs and limits of phenotypic plasticity. *Trends Ecol. Evol.* **1998**, *13*, 77–81. [CrossRef]

44. Michalski, S.; Malyshev, A.V.; Kreyling, J. Trait variation in response to varying winter temperatures, diversity patterns and signatures of selection along the latitudinal distribution of the widespread grasSLAnd plantArrhenatherum elatius. *Ecol. Evol.* **2017**, *7*, 3268–3280. [CrossRef] [PubMed]

45. Kreyling, J.; Puechmaille, S.J.; Malyshev, A.V.; Valladares, F. Phenotypic plasticity closely linked to climate at origin and resulting in increased mortality under warming and frost stress in a common grass. *Ecol. Evol.* **2019**, *9*, 1344–1352. [CrossRef] [PubMed]

46. Fick, S.E.; Hijmans, R.J. WorldClim 2: New 1-km spatial resolution climate surfaces for global land areas. *Int. J. Climatol.* **2017**, *37*, 4302–4315. [CrossRef]

47. Ellenberg, H. *Vegetation Ecology of Central Europe*; Cambridge University Press: Cambridge, UK; New York, NY, USA; New Rochelle, NY, USA; Melbourne, VIC, Australia; Sydney, NSW, Australia, 1988.

48. Führer, E.; Horváth, L.; Jagodics, A.; Machon, A.; Pödör, Z. Application of a new aridity index in Hungarian forestry practice. *Idojaras* **2011**, *3*, 205–216.

49. Fang, J.; Lechowicz, M.J. Climatic limits for the present distribution of beech (Fagus L.) species in the world. *J. Biogeogr.* **2006**, *33*, 1804–1819. [CrossRef]

50. Stojanović, D.B.; Kržič, A.; Matović, B.; Orlović, S.; Duputié, A.; Djurdjević, V.; Galic, Z.; Stojnić, S. Prediction of the European beech (Fagus sylvatica L.) xeric limit using a regional climate model: An example from southeast Europe. *Agric. For. Meteorol.* **2013**, *176*, 94–103. [CrossRef]

51. Gavrilov, M.B.; Lukić, T.; Janc, N.; Basarin, B.; Marković, S.B. Forestry Aridity Index in Vojvodina, North Serbia. *Open Geosci.* **2019**, *11*, 367–377. [CrossRef]

52. Von Wühlisch, G.; Liesebach, M.; Muhs, H.J.; Stephan, B.R. A network of international beech provenance trials. In *First EUFORGEN Meeting on Social Broadleaves*; Turok, J., Kremer, A., de Vries, S., Eds.; International Plant Genetic Resources Institute: Rome, Italy, 1998; pp. 164–172.

53. Vega, C.; González, G.; Bahamonde, H.A.; Valbuena-Carabaña, M.; Gil, L.; Fernández, V. Effect of irradiation and canopy position on anatomical and physiological features of Fagus sylvatica and Quercus petraea leaves. *Plant Physiol. Biochem.* **2020**, *152*, 232–242. [CrossRef] [PubMed]

54. Aranda, I.; Pardo, F.; Gil, L.; Pardos, J. Anatomical basis of the change in leaf mass per area and nitrogen investment with relative irradiance within the canopy of eight temperate tree species. *Acta Oecologica* **2004**, *25*, 187–195. [CrossRef]

55. Wolf, L. *Mikroskopicka Tehnica*; National Health Publishing: Praha, Czech Republic, 1950.

56. Holland, N.; Richardson, A.D. Stomatal Length Correlates with Elevation of Growth in Four Temperate Speciest. *J. Sustain. For.* **2009**, *28*, 63–73. [CrossRef]

57. Sack, L.; Cowan, P.D.; Jaikumar, N.; Holbrook, N.M. The "hydrology" of leaves: Co-ordination of structure and function in temperate woody species. *Plant Cell Environ.* **2003**, *26*, 1343–1356. [CrossRef]

58. Vile, D.; Garnier, É.; Shipley, B.; Laurent, G.; Navas, M.-L.; Roumet, C.; Lavorel, S.; Díaz, S.; Hodgson, J.G.; Lloret, F.; et al. Specific Leaf Area and Dry Matter Content Estimate Thickness in Laminar Leaves. *Ann. Bot.* **2005**, *96*, 1129–1136. [CrossRef]

59. Aranda, I.; Bergasa, L.F.; Gil, L. Effects of relative irradiance on the leaf structure of Fagus sylvatica L. seedlings planted in the understory of a Pinus sylvestris L. stand after thinning. *Ann. For. Sci.* **2001**, *58*, 673–680. [CrossRef]

60. Mátyás, C.; Fady, B.; Vendramin, G. Forests at the limit: Evolutionary—genetic consequences of environmental changes at the receding (xeric) edge of distribution. Report from a researcher workshop. *Acta Silv Lign Hung* **2009**, *66*, 201–204. [CrossRef]

61. Pinheiro, J.; Bates, D.; DebRoy, S.; Sarkar, D.; R Core Team. *nlme: Linear and Nonlinear Mixed Effects Models*; R package version 3.1-150; 2020; Available online: https://cran.r-project.org (accessed on 28 October 2020).

62. Bartoń, K. *MuMIn: Multi-Model Inference*; R package version 1.43.17; 2020; Available online: https://cran.r-project.org (accessed on 28 October 2020).

63. Nakagawa, S.; Schielzeth, H. A general and simple method for obtainingR2from generalized linear mixed-effects models. *Methods Ecol. Evol.* **2012**, *4*, 133–142. [CrossRef]

64. Marron, N.; Dreyer, E.; Boudouresque, E.; DeLay, D.; Petit, J.-M.; Delmotte, F.M.; Brignolas, F. Impact of successive drought and re-watering cycles on growth and specific leaf area of two Populus x canadensis (Moench) clones, "Dorskamp" and "Luisa_Avanzo". *Tree Physiol.* **2003**, *23*, 1225–1235. [CrossRef]

65. Robson, T.M.; Sánchez-Gómez, D.; Cano, F.J.; Aranda, I. Variation in functional leaf traits among beech provenances during a Spanish summer reflects the differences in their origin. *Tree Genet. Genomes* **2012**, *8*, 1111–1121. [CrossRef]

66. Meier, I.C.; Leuschner, C. Genotypic variation and phenotypic plasticity in the drought response of fine roots of European beech. *Tree Physiol.* **2008**, *28*, 297–309. [CrossRef] [PubMed]

67. Aspelmeier, S.; Leuschner, C. Genotypic variation in drought response of silver birch (Betula pendula Roth): Leaf and root morphology and carbon partitioning. *Trees* **2005**, *20*, 42–52. [CrossRef]

68. Dunn, J.; Hunt, L.; Afsharinafar, M.; Al Meselmani, M.; Mitchell, A.; Howells, R.; Wallington, E.; Fleming, A.J.; Gray, J.E. Reduced stomatal density in bread wheat leads to increased water-use efficiency. *J. Exp. Bot.* **2019**, *70*, 4737–4748. [CrossRef] [PubMed]

69. Lawson, T.; Matthews, J. Guard Cell Metabolism and Stomatal Function. *Annu. Rev. Plant Biol.* **2020**, *71*, 273–302. [CrossRef] [PubMed]

70. Bussotti, F.; Prancrazi, M.; Matteucci, G.; Gerosa, G. Leaf morphology and chemistry in Fagus sylvatica (beech) trees as affected by site factors and ozone: Results from CONECOFOR permanent monitoring plots in Italy. *Tree Physiol.* **2005**, *25*, 211–219. [CrossRef] [PubMed]

71. Stojnić, S.; Orlovic, S.; Trudić, B.; Živković, U.; Von Wuehlisch, G.; Miljkovic, D. Phenotypic plasticity of European beech (Fagus sylvatica L.) stomatal features under water deficit assessed in provenance trial. *Dendrobiology* **2015**, *73*, 163–173. [CrossRef]

72. Stojnić, S.; Orlović, S.; Miljković, D.; Galić, Z.; Kebert, M.; Von Wuehlisch, G. Provenance plasticity of European beech leaf traits under differing environmental conditions at two Serbian common garden sites. *Eur. J. For. Res.* **2015**, *134*, 1109–1125. [CrossRef]

73. Li, Y.; Li, H.; Li, Y.; Zhang, S. Improving water-use efficiency by decreasing stomatal conductance and transpiration rate to maintain higher ear photosynthetic rate in drought-resistant wheat. *Crop. J.* **2017**, *5*, 231–239. [CrossRef]

74. Lawson, T.; Blatt, M.R.; Yu, L.; Shi, D.; Li, J.; Kong, Y.; Yu, Y.; Chai, G.; Hu, R.; Wang, J.; et al. Stomatal Size, Speed, and Responsiveness Impact on Photosynthesis and Water Use Efficiency. *Plant Physiol.* **2014**, *164*, 1556–1570. [CrossRef]

75. Diazlopez, L.; Gimeno, V.; Simón, I.; Martínez, V.; Rodríguez-Ortega, W.; García-Sánchez, F. Jatropha curcas seedlings show a water conservation strategy under drought conditions based on decreasing leaf growth and stomatal conductance. *Agric. Water Manag.* **2012**, *105*, 48–56. [CrossRef]

76. George-Jaeggli, B.; Mortlock, M.Y.; Borrell, A.K. Bigger is not always better: Reducing leaf area helps stay-green sorghum use soil water more slowly. *Environ. Exp. Bot.* **2017**, *138*, 119–129. [CrossRef]

77. Anyia, A. Water-use efficiency, leaf area and leaf gas exchange of cowpeas under mid-season drought. *Eur. J. Agron.* **2004**, *20*, 327–339. [CrossRef]

78. Songsri, P.; Jogloy, S.; Holbrook, C.; Kesmala, T.; Vorasoot, N.; Akkasaeng, C.; Patanothai, A. Association of root, specific leaf area and SPAD chlorophyll meter reading to water use efficiency of peanut under different available soil water. *Agric. Water Manag.* **2009**, *96*, 790–798. [CrossRef]

79. Poorter, L.; Markesteijn, L. Seedling Traits Determine Drought Tolerance of Tropical Tree Species. *Biotropica* **2008**, *40*, 321–331. [CrossRef]

80. Slot, M.; Poorter, L. Diversity of Tropical Tree Seedling Responses to Drought. *Biotropica* **2007**, *39*, 683–690. [CrossRef]

81. Basu, S.; Ramegowda, V.; Kumar, A.; Pereira, A. Plant adaptation to drought stress. *F1000Research* **2016**, *5*, 1554. [CrossRef]

82. Tanaka, Y.; Sugano, S.S.; Shimada, T.; Hara-Nishimura, I. Enhancement of leaf photosynthetic capacity through increased stomatal density in Arabidopsis. *New Phytol.* **2013**, *198*, 757–764. [CrossRef]

83. Harrison, E.L.; Cubas, L.A.; Gray, J.E.; Hepworth, C. The influence of stomatal morphology and distribution on photosynthetic gas exchange. *Plant J.* **2020**, *101*, 768–779. [CrossRef]

84. Davis, M.A.; Wrage, K.J.; Reich, P.B.; Tjoelker, M.G.; Schaeffer, T.; Muermann, C. Survival, growth, and photosynthesis of tree seedlings competing with herbaceous vegetation along a water-light-nitrogen gradient. *Plant Ecol.* **1999**, *145*, 341–350. [CrossRef]

85. Franks, P.J.; Beerling, D.J. Maximum leaf conductance driven by CO_2 effects on stomatal size and density over geologic time. *Proc. Natl. Acad. Sci. USA* **2009**, *106*, 10343–10347. [CrossRef]

86. Fanourakis, D.; Briese, C.; Max, J.; Kleinen, S.; Putz, A.; Fiorani, F.; Ulbrich, A.; Schurr, U. Rapid determination of leaf area and plant height by using light curtain arrays in four species with contrasting shoot architecture. *Plant Methods* **2014**, *10*, 9. [CrossRef] [PubMed]

87. Vráblová, M.; Vrábl, D.; Hronková, M.; Kubásek, J.; Šantrůček, J. Stomatal function, density and pattern, and CO_2 assimilation in Arabidopsis thaliana tmm1 and sdd1-1 mutants. *Plant Biol.* **2017**, *19*, 689–701. [CrossRef] [PubMed]

88. Doheny-Adams, T.; Hunt, L.; Franks, P.J.; Beerling, D.J.; Gray, J.E. Genetic manipulation of stomatal density influences stomatal size, plant growth and tolerance to restricted water supply across a growth carbon dioxide gradient. *Philos. Trans. R. Soc. B Biol. Sci.* **2012**, *367*, 547–555. [CrossRef] [PubMed]

89. Bucher, S.F.; Auerswald, K.; Tautenhahn, S.; Geiger, A.; Otto, J.; Müller, A.; Römermann, C. Inter—and intraspecific variation in stomatal pore area index along elevational gradients and its relation to leaf functional traits. *Plant Ecol.* **2016**, *217*, 229–240. [CrossRef]

90. Brodribb, T.J.; Holbrook, N.M.; Zwieniecki, M.A.; Palma, B. Leaf hydraulic capacity in ferns, conifers and angiosperms: Impacts on photosynthetic maxima. *New Phytol.* **2004**, *165*, 839–846. [CrossRef] [PubMed]

91. Barigah, T.; Saugier, B.; Mousseau, M.; Guittet, J.; Ceulemans, R.J. Photosynthesis, leaf area and productivity of 5 poplar clones during their establishment year. *Ann. Sci. For.* **1994**, *51*, 613–625. [CrossRef]

92. Richards, R.A. Selectable traits to increase crop photosynthesis and yield of grain crops. *J. Exp. Bot.* **2000**, *51*, 447–458. [CrossRef]

93. Gómez-del-Campo, M.; Ruiz, C.; Lissarrague, J.R. Effect of Water Stress on Leaf Area Development, Photosynthesis, and Productivity in Chardonnay and Airén Grapevines. *Am. J. Enol. Vitic.* **2002**, *53*, 138–143.

94. Wortemann, R.; Herbette, S.; Barigah, T.S.; Fumanal, B.; Alia, R.; Ducousso, A.; Gomory, D.; Roeckel-Drevet, P.; Cochard, H. Genotypic variability and phenotypic plasticity of cavitation resistance in Fagus sylvatica L. across Europe. *Tree Physiol.* **2011**, *31*, 1175–1182. [CrossRef]

95. Čortan, D.; Nonić, M.; Šijačić-Nikolić, M. Phenotypic Plasticity of European Beech from International Provenance Trial in Serbia. In *Advances in Global Change Research*; Springer Science and Business Media LLC: Berlin/Heidelberg, Germany, 2019; pp. 333–351.

96. Petkova, K.; Molle, E.; Konnert, M.; Knutzen, F. Comparing German and Bulgarian provenances of European beech (Fagus sylvatica L.) regarding survival, growth and ecodistance. *Silva Balc.* **2019**, *2*, 27–48.

97. Sáenz-Romero, C.; Kremer, A.; Nagy, L.; Újvári-Jármay, É.; Ducousso, A.; Kóczán-Horváth, A.; Hansen, J.K.; Mátyás, C. Common garden comparisons confirm inherited differences in sensitivity to climate change between forest tree species. *Peer J.* **2019**, *7*, e6213. [CrossRef]

98. Robson, T.M.; Rasztovits, E.; Aphalo, P.J.; Alía, R.; Aranda, I. Flushing phenology and fitness of European beech (Fagus sylvatica L.) provenances from a trial in La Rioja, Spain, segregate according to their climate of origin. *Agric. For. Meteorol.* **2013**, *180*, 76–85. [CrossRef]

99. Frank, A.; Pluess, A.R.; Howe, G.T.; Sperisen, C.; Heiri, C. Quantitative genetic differentiation and phenotypic plasticity of European beech in a heterogeneous landscape: Indications for past climate adaptation. *Perspect. Plant Ecol. Evol. Syst.* **2017**, *26*, 1–13. [CrossRef]

100. Gianoli, E.; Valladares, F. Studying phenotypic plasticity: The advantages of a broad approach. *Biol. J. Linn. Soc.* **2011**, *105*, 1–7. [CrossRef]

101. Hewitt, G.M. Some genetic consequences of ice ages, and their role in divergence and speciation. *Biol. J. Linn. Soc.* **1996**, *58*, 247–276. [CrossRef]

102. Hewitt, G.M. Post-glacial re-colonization of European biota. *Biol. J. Linn. Soc.* **1999**, *68*, 87–112. [CrossRef]

103. Lander, T.A.; Klein, E.K.; Roig, A.; Oddou-Muratorio, S. Weak founder effects but significant spatial genetic imprint of recent contraction and expansion of European beech populations. *Heredity* **2020**, 1–14. [CrossRef]

104. Callahan, H.S.; Maughan, H.; Steiner, U.K. Phenotypic Plasticity, Costs of Phenotypes, and Costs of Plasticity. *Ann. N. Y. Acad. Sci.* **2008**, *1133*, 44–66. [CrossRef]

105. McDonald, J.M.C.; Ghosh, S.M.; Gascoigne, S.J.L.; Shingleton, A.W. Plasticity Through Canalization: The Contrasting Effect of Temperature on Trait Size and Growth in Drosophila. *Front. Cell Dev. Biol.* **2018**, *6*, 156. [CrossRef]

106. Debat, V.; David, P. Mapping phenotypes: Canalization, plasticity and developmental stability. *Trends Ecol. Evol.* **2001**, *16*, 555–561. [CrossRef]

107. Klingenberg, C.P. Phenotypic Plasticity, Developmental Instability, and Robustness: The Concepts and How They Are Connected. *Front. Ecol. Evol.* **2019**, *7*, 7. [CrossRef]

108. Yeh, P.J.; Price, T.D. Adaptive Phenotypic Plasticity and the Successful Colonization of a Novel Environment. *Am. Nat.* **2004**, *164*, 531–542. [CrossRef] [PubMed]

109. Nicotra, A.B.; Atkin, O.; Bonser, S.; Davidson, A.; Finnegan, E.; Mathesius, U.; Poot, P.; Purugganan, M.; Richards, C.; Valladares, F.; et al. Plant phenotypic plasticity in a changing climate. *Trends Plant Sci.* **2010**, *15*, 684–692. [CrossRef] [PubMed]

110. Ghalambor, C.K.; McKay, J.K.; Carroll, S.P.; Reznick, D.N. Adaptive versus non-adaptive phenotypic plasticity and the potential for contemporary adaptation in new environments. *Funct. Ecol.* **2007**, *21*, 394–407. [CrossRef]

111. Anderson, J.T.; Song, B. Plant adaptation to climate change—Where are we? *J. Syst. Evol.* **2020**, *58*, 533–545. [CrossRef]

112. Van Buskirk, J.; Steiner, U.K. The fitness costs of developmental canalization and plasticity. *J. Evol. Biol.* **2009**, *22*, 852–860. [CrossRef]

113. Scheiner, S.M. The genetics of phenotypic plasticity. XVI. Interactions among traits and the flow of information. *Evolution* **2018**, *72*, 2292–2307. [CrossRef]

Publisher's Note: MDPI stays neutral with regard to jurisdictional claims in published maps and institutional affiliations.

 *forests*

Article

Tree Line Shift in the Olympus Mountain (Greece) and Climate Change

Athanasios Zindros, Kalliopi Radoglou, Elias Milios and Kyriaki Kitikidou *

Department of Forestry and Management of the Environment and Natural Resources, Democritus University, Pandazidou 193, 68200 Orestiada, Greece; szintros@yahoo.com (A.Z.); kradoglo@fmenr.duth.gr (K.R.); emilios@fmenr.duth.gr (E.M.)
* Correspondence: kkitikid@fmenr.duth.gr

Received: 11 August 2020; Accepted: 10 September 2020; Published: 14 September 2020

Abstract: One of the effects of climate change is, among others, changes to forest ecosystems. Research Highlights: Temperature increases and upward tree line shifts are linked in many studies. However, the impact of climate change on tree lines has not been studied in Greece. Background and Objectives: The aim of this study is to assess the relation of tree line shifts and climate change in Olympus mountain, and especially in a protected area. Materials and Methods: In the Olympus mountain, which includes a protected area (the Olympus National Park core) since 1938, GIS data regarding forest cover were analyzed, while climate change from a previous study is presented. Results: Forest expansion and an upward tree line shift are proven in the Olympus mountain area. In the National Park core, the tree line shift is the result of climate change and attributed to the significant temperature increase in the growing season. Conclusions: There are strong indications that a temperature increase leads to an upward shift of the tree lines in the National Park core.

Keywords: climate change; National Park; tree line shift

---

1. Introduction

The natural environment is expected to be strongly impacted by global warming. Wide parts of Southern Europe will be affected severely by climate impacts [1]. One of the major effects of climate change on forest ecosystem is changes in forest boundaries, with upward shifts in tree lines [2–5]. According to Paulsen et al. [6], there is an abrupt shift in the growth rates of the tree line, especially in plant species such as Alpine [6]. As Motta and Nola [7] postulate, there are significant growth trends and changing aspects of the alpine plant communities and species compositions [7]. Young trees are known for responding to the environmental change and their establishment in forest gaps close to the tree line may be seen in many areas [8–10]. However, it is important to understand the drivers of these trends in the alpine ecological community [11].

Firstly, these trends could be a result of a significant shift in climate change. Due to the temperature changes, alpine trees are directly influenced; therefore, it is anticipated to lead to structural shifts of the composition of trees and in a rise of the alpine tree line [7].

A probable tree line shift of around 200 m is expected to be found in the Swiss Alps as an effect of changes in the climate that have been experienced since 1980 [12,13]. The shift can be calculated by looking at the difference in temperatures between altitudes—a linear reduction in temperature of 0.55 °C per 100 m elevation was recorded. Walther et al. [10] discovered that the vegetation in the southeastern Swiss Alps had also experienced constant change since 1985, and concluded that the change in vegetation was resulted by an increased temperature at the related sites.

This research will entail an examination to determine the connection of shift tree lines and climate change. This examination will be conducted by answering the following questions: 1. Is there a tree line shift in Olympus mountain? 2. Is there a significant temperature or/and precipitation variability in

Olympus mountain? Positive responses to the above can justify the dependence of tree line shifts on climate change in the protected area of the mountain (i.e., the National Park core), given that this area was declared a National Park in 1938 (the core of the National Park is approximately 4000 ha) and, at least since 1981 [14], no management was implemented.

2. Materials and Methods

2.1. Study Area

The study area is the central part of Olympus mountain that was bare of vegetation (meadow) in 1960. Apart from its historical value and importance in Greek mythology, which are recognized not only in Greece but also in Europe and worldwide, Olympus mountain is also noted for its ecological value, regarding its exceptional biodiversity and rich flora.

Olympus mountain is the highest mountain in Greece, with varying elevations. Its type of climate, according to the Köppen climate classification, is Cfb = temperate oceanic climate, meaning that the coldest month averages above 0 or −3 °C, and all months have average temperatures below 22 °C, and at least four months average above 10 °C. There is no significant precipitation difference between seasons. This climate description of the study area is based on meteorological data of the nearest meteorological station to the study area, Litochoro (coordinates, 40.091 N, 22.361 E; elevation 2579 m; climate data for the period 1901–2018) [15]. These data are illustrated in a climate chart (Figure 1).

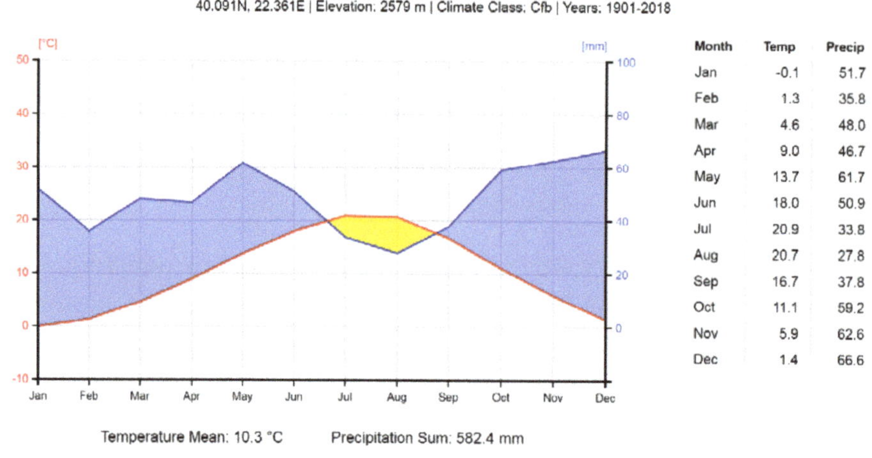

Month	Temp	Precip
Jan	-0.1	51.7
Feb	1.3	35.8
Mar	4.6	48.0
Apr	9.0	46.7
May	13.7	61.7
Jun	18.0	50.9
Jul	20.9	33.8
Aug	20.7	27.8
Sep	16.7	37.8
Oct	11.1	59.2
Nov	5.9	62.6
Dec	1.4	66.6

Figure 1. Ombrothermic diagram of Litochoro [15].

Today the Olympus National Park Management Agency manage a larger area than the core area of the National Park. This area has ecological importance, due to the rare and useful flora and fauna environment. The area also carries geomorphological formations and stream of waters. The area is protected through a legal arrangement that ensures that the park maintains its ecological friendly environment. Furthermore, the park became the first area to be regulated by a Royal Decree; the Decree was issued in 1938. The law was, however, enacted in 1985, when the park became fully operational. The area has been classified by the United Nations Educational, Scientific and Cultural Organization (UNESCO) as a biosphere reserve since 1981. The area offers a natural environment for the conservation of wild birds, and is protected by the NATURA 2000 Network program, which is certified by the habitats directives for wild birds [16].

Olympus consists mainly of dolomite limestones and marbles of various dimensions and ages. Gneisses are found on the western slopes of Mount Olympus, and flysch occurs locally. The predominance of limestone significantly affects the climate and the appearance of vegetation [17].

2.2. Forest Cover and Tree Line Shift

Forest expansion in the study area was quantified using data from the Hellenic Military Geographical Service, for the year 1960 [18], and the National Cadastre and Mapping Agency S.A. company, for the year 2020 [19]. Aerial photographs of the year 1960 were purchased by the Hellenic Military Geographical Service, along with a license. For the year 2020, mosaics of orthophotographs were used, which were purchased from the National Cadastre and Mapping Agency S.A. company [19]. Geographic data from the ASTER GDEM v2 Worlwide Elevation Data (1 arc-second Resolution) were used [20]. Data were analyzed by using QGIS (free software) [21] and Global Mapper (partially licensed software) [22].

There are many terms regarding the shift of vegetation covered areas to bare areas [23]. In this study, in both measurements (for the years 1960 and 2020), the tree line was considered to be the line created by the limits of closed forest.

2.3. Climate Change

The assumption that the climate has really changed in the study area is based entirely on the work of Klesse et al. [24]. The dendrochronological investigation of black pine (*Pinus nigra* Arn.) from the Olympus mountain, extending from 850 to 1700 m in elevation, conducted by Klesse et al. [24] using the standard dendrochronological technique of Stokes and Smiley [25], showed significant long-term deviations in temperature and precipitation over time, with summers tending to be significantly warmer (late summer) and wetter (early summer) than the summers of the last century.

3. Results–Discussion

3.1. Forest Cover and Tree Line Shift

During the last 60 years, Olympus mountain lost more than one quarter of its area bare of forest vegetation. The study area that we have digitized for 1960 was calculated to have an area equal to 126.536 km^2; this was area bare of forest vegetation (including alpine mountain meadow). The area bare of forest vegetation, bounded by the tree line, in the year 2020, was calculated equal to 88.185 km^2 (Figure 2).

More analytically, in the east side there is a decrease in the area bare of forest vegetation, i.e., an increase in the forested area, by 7.088 km^2, in the west side by 16,637 km^2, in the south side by 9.688 km^2, and in the north side by 4.937 km^2. In the core of the National park, the increase in the forested area is 3.837 km^2. Besides forest expansion (increase by 38.351 km^2), in Figure 2 we observe that, at some points the new tree lines touch the old ones, while they have a general upwards trend. However, in a very few spots (mainly in north side), the forest retreated slightly in 2020 compared to 1960.

In the east side, the lowest altitude of the tree line in 1960 was 1848 m, while in 2020 in that particular area the altitude of the tree line goes up to 2082 m. On the same side, in 2020, the highest altitude of the tree line was 2496 m, while in 1960 that particular area was 2138 m. On the west side the lowest altitude of the tree line in 1960 was 1132 m, while in 2020 that particular area goes up to 1629 m. On the same side, in 2020, the highest altitude of the tree line was 2230 m, while in 1960 that particular area was 1595 m. On the north side, the lowest altitude of the tree line in 1960 was 1568 m, while in 2020 that particular area goes up to 1800 m. On the same side, in 2020, the highest altitude of the tree line was 2000 m, while in 1960 that particular area was 1750 m. On the south side, the lowest altitude of the tree line in 1960 was 1285 m, while in 2020 that particular area goes up to 1661 m. On the same side, in 2020, the highest altitude of the tree line was 2230 m, while in 1960 that particular area was

1595 m. The core of the National Park (declared in 1938) exhibits the tree line altitudes of the east side, since it is incorporated in that side and the described altitudes appear in the core.

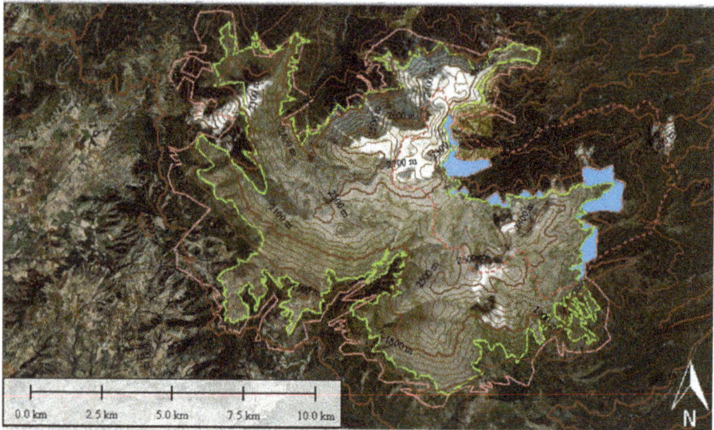

Figure 2. Area bare of vegetation in the year 1960 (vivid tangerine red line), and in the year 2020 (chartreuse green line) with contour lines per 100 m (reference: the year 2020). Area bare of vegetation (decrease) in the core of the National Park (sky-blue color).

3.2. Climate Change and Land Use Changes

According to the results of this study, the highest tree line altitude in Olympus is 2496 m. According to Theodoropoulos et al. [26] the forest in Olympus approximately reaches the altitude of 2500 m.

A tree line ascent involves changes in a number of consecutive processes (viable seed production, seed spreading, appropriate regeneration sites, germination, seedlings survival and growth); consequently, changes in temperature and precipitation alone does not unavoidably mean that there will be a tree line shift [27,28]. Even with the tree line shift, the climate change relationship is very complex, on a global scale, and one of the main factors that affects tree lines is the temperature of the growing season [29,30]. Forest expansion with an upward shift of the tree line is linked to a temperature increase [10,31–35]; however, changes in land use have a great impact on the upward shift of the tree line [12,36]. Human activities such as the grazing of domesticated animals affect tree mountain lines [37]. Moreover, Speed et al. [38] mention that, along with climate change, changes of land use have to be considered as factors affecting tree line ecotone shifts. Grazing strongly affects forests in Greece [39–41]. Near the west side of the study area there are villages where the grazing of domesticated animals was one of the main land uses [17,42]. Grazing is a significant disturbance factor in the area of the Olympus mountain [17,26]. The movement of the tree line at higher altitudes in almost the entire study area, and, especially in the west side, is strongly related to the gradual abandonment of the traditional land use. This is obvious from the altitude where the area bare from vegetation is observed in most areas in 1960 and in 2020. The abandonment of grazing and other land uses is a crucial factor in the formation of structures and in the development of patterns of forests in mountainous areas [43,44]. However, it seems that, in the National Park core, the tree line shift is, actually, the result of climate change. Given the fact that there was a significant change in climate in the area, this conclusion is based on the following facts: (a) there were no villages and consequently there were no human-induced disturbances in the vicinity of the core, (b) human activities such as grazing have been forbidden since 1981 [14]—most likely grazing had stopped many years before 1981, due to the protection status of the core—(c) the steep ground slopes of many areas of the core make the area unsuitable for gazing and (d) the elevation of the tree line changes between 1960 and 2020.

The shift in the tree line in the entire study area (not just the National Park core) is significant, and, even though the study area is not protected (except for the National Park), climate change probably accelerated the tree line shift in general. The influences of climate change and the reduction in or the elimination of grazing are difficult to study separately [37,45]. Unfortunately, there are no available livestock data for the study area in order to assess the impact of grazing on the tree line shift. The only official livestock data are those of the Greek Payment Authority of Common Agricultural Policy (CAP) Aid Schemes (url: http://aggregate.opekepe.gr), and are available only for the East Olympos, and only for the period 2011–2014 (after 2014 the neighboring municipalities were merged; consequently, livestock data were aggregated for a number of municipalities and are not comparable with those before 2014). Nevertheless, since there was no management implemented in the Olympus National Park core (absence of grazing for many decades), we can claim that the cause of the tree line shift in the National Park core is, exclusively, climate change.

4. Conclusions

An increase in forest cover by 38.351 km^2 with an upward shift of the tree line from 1960 to 2020, has been noted in Olympus mountain, while the corresponding area for the Olympus National Park core is 3.837 km^2. The movement of the tree line at higher altitudes in almost the entire study area, and especially on the west side, is related to the gradual abandonment of the traditional land use, i.e., grazing. However, in the Olympus National Park core, the tree line shift is solely the result of climate change in the area, and more specifically because of warmer summers. This conclusion is based mainly on the protection status of the core (absence of grazing for many decades), combined with other factors such as altitude changes of the tree line, and steep slopes.

The forested areas are likely to increase further, if the climate keeps getting warmer. Future research should focus on long-term and area-wide observations of regeneration and tree growth. It should also catch more factors that affect the tree line ecotone and recognize any hidden facilitators.

Author Contributions: Conceptualization, A.Z., K.R. Data curation, A.Z., K.R. Methodology, A.Z., K.R., K.K., E.M. Writing-original draft preparation, A.Z., K.R., K.K., E.M. All authors have read and agreed to the published version of the manuscript.

Funding: This research was funded by the LIFE project FoResMit "Recovery of degraded coniferous Forests for environmental sustainability Restoration and climate change Mitigation" grant number [LIFE14 CCM/IT/000905]

Acknowledgments: We would like to express our great appreciation to the Hellenic Military Geographical Service for the mosaics of orthophotographs processing. Our special thanks to the Olympus National Park Management Agency for their assistance regarding the study area description.

Conflicts of Interest: The authors declare no conflict of interest.

References

1. IPCC: Summary for Policymakers. In *Climate Change 2013: The Physical Science Basis. Contribution of Working Group I to the Fifth Assessment Report of the Intergovernmental Panel on Climate Change*; Stocker, T.F., Qin, D., Plattner, G.-K., Tignor, M.M.B., Allen, S.K., Boschung, J., Nauels, A., Xia, Y., Bex, V., Midgley, P.M., Eds.; Cambridge University Press: New York, NY, USA, 2013.

2. Cudlín, P.; Klopčič, M.; Tognetti, R.; Máli&, F.; Alados, C.L.; Bebi, P.; Grunewald, K.; Zhiyanski, M.; Andonowski, V.; La Porta, N.; et al. Drivers of treeline shift in different European mountains. *Clim. Res.* **2017**, *73*, 135–150. [CrossRef]

3. Mainali, K.; Shrestha, B.; Sharma, R.; Adhikari, A.; Gurarie, E.; Singer, M.; Parmesan, C. Contrasting responses to climate change at Himalayan treelines revealed by population demographics of two dominant species. *Ecol. Evol.* **2020**, *10*, 1209–1222. [CrossRef] [PubMed]

4. Naccarella, A.; Morgan, J.; Cutler, S.; Venn, S. Alpine treeline ecotone stasis in the face of recent climate change and disturbance by fire. *PLoS ONE* **2020**, *15*, e0231339. [CrossRef]

5. Lett, S.; Dorrepaal, E. Global drivers of tree seedling establishment at alpine treelines in a changing climate. *Funct. Ecol.* **2018**, *32*, 1666–1680. [CrossRef]

6. Paulsen, J.; Weber, U.M.; Körner, C. Tree growth near tree line: Abrupt or gradual reduction with altitude? *Arct. Antarct. Alp. Res.* **2000**, *32*, 14–20. [CrossRef]

7. Motta, R.; Nola, P. Growth trends and dynamics in subalpine forest stands in the Varaita Valley (Piedmont, Italy) andtheir relationships with human activities and global change. *J. Veg. Sci.* **2001**, *12*, 219–230. [CrossRef]

8. Keller, F.; Kienast, F.; Beniston, M. Evidence of response of vegetation to environmental change on high-elevation sites in the Swiss Alps. *Reg. Environ. Chang.* **2000**, *1*, 70–77. [CrossRef]

9. Pauli, H.; Gottfried, M.; Grabherr, G. High summits of the Alps in a changing climate. The oldest observation series on high mountain plant diversity in Europe, 2001. In *Fingerprints of Climate Change. Adapted Behaviour and Shifting Species Ranges*; Walther, G.R., Burga, C.A., Edwards, P.J., Eds.; Springer: Boston, MA, USA, 2001; pp. 139–149.

10. Walther, G.; Beißner, S.; Burga, C. Trends in the upward shift of alpine plants. *J. Veg. Sci.* **2005**, *16*, 541–548. [CrossRef]

11. Körner, C. *Alpine Plant Life-Functional Plant Ecology of High Mountain Ecosystems*, 2nd ed.; Springer: Berlin/Heidelberg, Germany, 2000.

12. Gehrig-Fasel, J.; Guisan, A.; Zimmermann, N. Tree line shifts in the Swiss Alps: Climate change or land abandonment? *J. Veg. Sci.* **2007**, *18*, 571–582. [CrossRef]

13. Jones, P.D.; Moberg, A. Hemispheric and large-scale surface air temperature variations: An extensive revision and an update to 2001. *J. Clim.* **2003**, *16*, 206–223. [CrossRef]

14. Mount Olympus—United Nations Educational, Scientific and Cultural Organization. Available online: http://www.unesco.org/new/en/natural-sciences/environment/ecological-sciences/biosphere-reserves/europe-north-america/greece/mount-olympus/ (accessed on 5 May 2020).

15. Harris, I.C.; Jones, P.D.; Osborn, T. CRU TS4.04: Climatic Research Unit (CRU) Time-Series (TS) version 4.04 of high-resolution gridded data of month-by-month variation in climate (January 1901–December 2019). *Cent. Environ. Data Anal.* **2020**, *25*. Available online: https://catalogue.ceda.ac.uk/uuid/89e1e34ec3554dc98594a5732622bce9/ (accessed on 5 May 2020).

16. Olympus National Park Management Agency. Available online: https://olympusfd.gr/ (accessed on 5 May 2020).

17. Bitos, I. The National Park of Mt Olympus in the International Status of Protected Areas—Problems of Its Management. MSc Thesis, Aristotle University of Thessaloniki, Thessaloniki, Serres, Greece, 2009. (In Greek).

18. Hellenic Military Geographical Service. Available online: http://web.gys.gr/portal/ (accessed on 5 May 2020).

19. The National Cadastre & Mapping Agency S.A. Available online: https://www.ktimatologio.gr/ (accessed on 5 May 2020).

20. ASTER Global Digital Elevation Map. Available online: https://asterweb.jpl.nasa.gov/gdem.asp (accessed on 5 May 2020).

21. Welcome to the QGIS Project! Available online: https://www.qgis.org/en/site/ (accessed on 5 May 2020).

22. Global Mapper-All-in-One GIS Software. Available online: https://www.bluemarblegeo.com/global-mapper/ (accessed on 5 May 2020).

23. Holtmeier, F.K. Mountain Timberlines: Ecology, Patchiness, and Dynamics. *Adv. Glob. Chang. Res.* **2009**, *36*, 11–28.

24. Klesse, S.; Ziehmer, M.; Rousakis, G.; Trouet, V.; Frank, D. Synoptic drivers of 400 years of summer temperature and precipitation variability on Mt. Olympus, Greece. *Clim. Dyn.* **2014**, *45*, 807–824. [CrossRef]

25. Stokes, M.A.; Smiley, T. *An Introduction to Tree-Ring Dating*; University of Chicago Press: Chicago, IL, USA, 1968.

26. Theodoropoulos, K.; Xystrakis, F.; Eleftheriadou, E.; Samaras, D. Vegetation Zones and Habitat Types in the Area of Responsibility of the Management Agency of Olympus National Park [In Greek with English Abstract]. In *Scientific Annals of the Faculty of Forestry and Natural Environment*; ME/2002/45; Aristotle University: Thessaloniki, Greece, 2011.

27. Kullman, L. Tree line population monitoring of *Pinus sylvestris* in the Swedish Scandes, 1973–2005: Implications for tree line theory and climate change ecology. *J. Ecol.* **2007**, *95*, 41–52. [CrossRef]

28. Wang, T.; Zhang, Q.-B.; Ma, K. Tree line dynamics in relation to climatic variability in the central Tianshan Mountains, northwestern China. *Glob. Ecol. Biogeogr.* **2006**, *15*, 406–415. [CrossRef]

29. Harsch, M.A.; Hulme, P.E.; McGlone, M.S.; Duncan, R.P. Are treelines advancing? A global meta-analysis of treeline response to climate warming. *Ecol. Lett.* **2009**, *12*, 1040–1049. [CrossRef]

30. Körner, C.; Paulsen, J. A world-wide study of high altitude treeline temperatures. *J. Biogeogr.* **2004**, *31*, 713–732. [CrossRef]

31. Banerjee, K.; Gatti, R.; Mitra, A. Climate change-induced salinity variation impacts on a stenoecious mangrove species in the Indian Sundarbans. *Ambio* **2016**, *46*, 492–499. [CrossRef]

32. Beckage, B.; Osborne, B.; Gavin, D.; Pucko, C.; Siccama, T.; Perkins, T. A rapid upward shift of a forest ecotone during 40 years of warming in the Green Mountains of Vermont. *Proc. Natl. Acad. Sci. USA* **2008**, *105*, 4197–4202. [CrossRef]

33. Gatti, R.C.; Callaghan, T.; Velichevskaya, A.; Dudko, A.; Fabbio, L.; Battipaglia, G.; Liang, J. Accelerating upward treeline shift in the Altai Mountains under last century climate change. *Sci. Rep.* **2019**, *9*, 7678. [CrossRef]

34. Liang, E.; Wang, Y.; Eckstein, D.; Luo, T. Little change in the fir tree-line position on the southeastern Tibetan Plateau after 200 years of warming. *New Phytol.* **2011**, *190*, 760–769. [CrossRef] [PubMed]

35. Walther, G.; Post, E.; Convey, P.; Menzel, A.; Parmesan, C.; Beebee, T.; Fromentin, J.M.; Hoegh-Guldberg, O.; Bairlein, F. Ecological responses to recent climate change. *Nature* **2002**, *416*, 389–395. [CrossRef] [PubMed]

36. Peñuelas, J.; Boada, M. A global change-induced biome shift in the Montseny mountains (NE Spain). *Glob. Chang. Biol.* **2003**, *9*, 131–140. [CrossRef]

37. Aas, B. Climatically Raised Birch Lines in Southeastern Norway 1918–1968. *Nor. Geogr. Tidsskr.* **1969**, *23*, 119–130. [CrossRef]

38. Speed, J.; Austrheim, G.; Hester, A.; Mysterud, A. Experimental evidence for herbivore limitation of the treeline. *Ecology* **2010**, *91*, 3414–3420. [CrossRef]

39. Bergmeier, E.; Dimopoulos, P. Identifying plant communities of thermophilous deciduous forest in Greece: Species composition, distribution, ecology and syntaxonomy. *Plant Biosyst.* **2008**, *142*, 228–254. [CrossRef]

40. Milios, E.; Pipinis, E.; Petrou, P.; Akritidou, S.; Smiris, P.; Aslanidou, M. Structure and regeneration patterns of the *Juniperus excelsa* Bieb. stands in the central part of the Nestos valley in the northeast of Greece, in the context of anthropogenic disturbances and nurse plant facilitation. *Ecol. Res.* **2007**, *22*, 713–723. [CrossRef]

41. Milios, E.; Pipinis, E.; Kitikidou, K.; Batziou, M.; Chatzakis, S.; Akritidou, S. Are sprouts the dominant form of regeneration in a lowland *Quercus pubescens—Quercus frainetto* remnant forest in Northeastern Greece? A regeneration analysis in the context of grazing. *New For.* **2014**, *45*, 165–177. [CrossRef]

42. Nezis, N. *Olympus (Geography-Nature-Culture-Tour-Mountaineering-Climbing-Place Names-Bibliography)*; Anavasis Publications: Athens, Greece, 2003. (In Greek)

43. Garbarino, M.; Lingua, E.; Weisberg, P.; Bottero, A.; Meloni, F.; Motta, R. Land-use history and topographic gradients as driving factors of subalpine *Larix decidua* forests. *Landsc. Ecol.* **2012**, *28*, 805–817. [CrossRef]

44. Milios, E. Dynamics and development patterns of *Pinus sylvestris* L.-*Fagus sylvatica* L. stands in central Rhodope. *Silva Gandav.* **2000**, *65*, 154–172. [CrossRef]

45. Bryn, A.; Potthoff, K. Elevational treeline and forest line dynamics in Norwegian mountain areas—A review. *Landsc. Ecol.* **2018**, *33*, 1225–1245. [CrossRef]

Article

High Nitrate or Ammonium Applications Alleviated Photosynthetic Decline of *Phoebe bournei* Seedlings under Elevated Carbon Dioxide

Xiao Wang [1], Xiaoli Wei [1,2,*], Gaoyin Wu [1] and Shenqun Chen [1]

[1]   College of Forestry, Guizhou University, Guiyang 550025, China; hopewangxiao@163.com (X.W.); wugaoyin1234@163.com (G.W.); squnchen@163.com (S.C.)

[2]   Institute for Forest Resources and Environment of Guizhou, Guizhou University, Guiyang 550025, China

*   Correspondence: gdwxl-69@126.com; Tel.: +86-139-8511-6043

Received: 14 February 2020; Accepted: 4 March 2020; Published: 6 March 2020

Abstract: *Phoebe bournei* is a precioustimber species and is listed as a national secondary protection plant in China. However, seedlings show obvious photosynthetic declinewhen grown long-term under an elevated CO_2 concentration (eCO_2). The global CO_2 concentration is predicted to reach 700 $\mu mol \cdot mol^{-1}$ by the end of this century; however, little is known about what causes the photosynthetic decline of *P. bournei* seedlings under eCO_2 or whether this photosynthetic decline could be controlled by fertilization measures. To explore this problem, one-year-old *P. bournei* seedlings were grown in an open-top air chamber under either an ambient CO_2 (aCO_2) concentration (350 ± 70 $\mu mol \cdot mol^{-1}$) or an eCO_2 concentration (700 ± 10 $\mu mol \cdot mol^{-1}$) from June 12th to September 8th and cultivated in soil treated with either moderate (0.8 g per seedling) or high applications (1.2 g per seedling) of nitrate or ammonium. Under eCO_2, the net photosynthetic rate (Pn) of *P. bournei* seedlings treated with a moderate nitrate application was 27.0% lower than that of seedlings grown under an aCO_2 concentration ($p < 0.05$), and photosynthetic declineappeared to be accompanied by a reduction of the electron transport rate (ETR), actual photochemical efficiency, chlorophyll content, ribulose-1,5-bisphosphate carboxylase/oxygenase (rubisco), rubisco activase (RCA) content, leaf thickness, and stomatal density. The Pn of seedlings treated with a high application of nitrate under eCO_2 was 5.0% lower than that of seedlings grown under aCO_2 ($p > 0.05$), and photosynthetic declineoccurred more slowly, accompanied by a significant increase in rubisco content, RCA content, and stomatal density. The Pn of *P. bournei* seedlings treated with either a moderate or a high application of ammonium and grown under eCO_2 was not significantly differentto that of seedlings grown under aCO_2—there was no photosynthetic decline—and the ETR, chlorophyll content, rubisco content, RCA content, and leaf thickness values were all increased. Increasing the application of nitrate or the supply of ammonium could slow down or prevent the photosynthetic declineof *P. bournei* seedlings under eCO_2 by changing the leaf structure and photosynthetic physiological characteristics.

Keywords: *Phoebe bournei*; nitrogen; carbon dioxide; photosynthesis; leaf anatomy

---

1. Introduction

A range of human activities such as fossil fuel combustion, deforestation, and industrial development are contributing to the rise of global CO_2 concentration [1]. CO_2, which is one of the main greenhouse gases in Earth's atmosphere, has increased in concentration from 280 $\mu mol \cdot mol^{-1}$ before the industrial revolution to more than 400 $\mu mol \cdot mol^{-1}$ at present. CO_2 concentration is predicted to reach 700 $\mu mol \cdot mol^{-1}$ by the end of this century, which is causing concern among the international community [2–4]. An elevated CO_2 (eCO_2) concentration not only increases the greenhouse effect of this gas but also affects other environmental factors on Earth's surface and directly or indirectly influences plant growth and physiology and various other aspects [5–8]. eCO$_2$ promoted photosynthesis of *Eucalyptus globules* [9] and *Coffea* spp. [10] and increased the content of non-structural carbohydrates of *Prunuspersica* [11].

Photosynthesis is an important physiological process that is necessary for plant growth and carbon sink function. Photosynthesis uses CO$_2$ as a reaction material when converting light energy into chemical energy, and the efficiency of these reactions are affected by changes in CO$_2$ concentration [12]. Given that CO$_2$ is one of the raw materials involved in the conversion of light energy into chemical energy, eCO$_2$ will affect the photosynthetic efficiency and other physiological activities of plants. Previous studies have characterized the short-term and long-term effects of eCO$_2$ concentrations on plant photosynthesis [13,14]. Rates of C3 plants (3-phosphoglycerate is the initial product of CO$_2$ fixation) photosynthesis are not saturated at current CO$_2$ levels and, therefore, eCO$_2$ concentrations will naturally stimulate plants to increase their photosynthetic rate and yield in the short-term [13]. In most cases, if long-term photosynthesis regulation does not occur, the photosynthetic rate will improve, accompanied by an increase in CO$_2$ concentration. This is mainly because the CO$_2$ concentration in intercellular spaces increases, directly increasing the amount of substrate for photosynthesis, which enhances the competitiveness and carboxylation efficiency of the ribulose-1,5-bisphosphate carboxylase/oxygenase (rubisco) enzyme, resulting in an increase in the net photosynthetic rate [15,16]. Studies on *Eucalyptus globules* [9] and *Coffea* spp. [10] have shown that short-term eCO$_2$ improves the net photosynthetic rate of these plants. However, the responses of plants that have been subjected to long-term eCO$_2$ treatment show greater variability. Under long-term eCO$_2$, the initial photosynthetic rate enhancement of various plants to eCO$_2$ gradually weakens or gradually disappears and, in some plants, photosynthesis is even downregulated [17–19]. This phenomenon has been reported by numerous studies and many studies have been performed to investigate why photosynthetic acclimation occurs [20,21]. One important reason is that a lack of a nitrogen supply impairs the C:N balance [19,22].

Nitrogen is an important component of plant proteins, nucleic acids, chlorophyll and some hormones and has obvious effects on photosynthesis and chlorophyll fluorescenceand, hence, is one of the essential nutrients in plants. Nitrogen nutrition influences the plant photosynthetic capacity through the decrease of synthesis of severalkey photosynthetic enzymes, especially of rubisco, thusaffecting carbon assimilation, and subsequently alsothe photochemical processes in thylakoid membranes [23]. The application of nitrogen is an important regulatory measure to improve plant photosynthetic characteristics and to promote plant growth [24]. For example, nitrogen regulates plant responses to different light intensities [25]. However, the effect of nitrogen nutrition on the adaptation of photosynthesis under eCO$_2$ depends on the supply level and is regulated by the source–sink relationship of plants [26,27]. An insufficient supply of nitrogen limits the response of plants to eCO$_2$ concentrations [26,28]. When nitrogen is the limiting factor, the nitrogen in rubisco may be redistributed to other enzymes or tissues and organs, which affects the synthesis of rubisco and, hence, reduces plant photosynthesis [29]. However, previous studies have indicated that eCO$_2$ may limit plant uptake of nitrogen and thus affect the positive growth response of plants to eCO$_2$ [30,31]. Furthermore, the form of nitrogen can also affect photosynthetic characteristics and the metabolism of different nitrogen forms is different. Most studies have shown that eCO$_2$ is less conducive for the absorption and transformation of nitrate by plants compared with that of ammonium [32–34].

Phoebe bournei (Hemsl.) Y.C. Yang is an evergreen tree, which is a precious timber species and an excellent ornamental species in China. The exposure of one-year-old *P. bournei* seedlings to eCO$_2$ for a short period has been shown to significantly increase their photosynthetic efficiency; however several months of treatment significantly inhibits photosynthesis, photosynthetic decline occurs, and the weakening of photosynthesis causes the metabolic level to decline [35]. In order to find a solution to the problem of photosynthetic decline under long-term eCO$_2$ conditions, such as those predicted to occur by the end of this century, the goals of this study were to understand whether different nitrogen forms and concentrations can alleviate the photosynthetic decline of *P. bournei* seedlings under eCO$_2$ and to elucidate the mechanism behind these traits. To investigate this problem, we grew one-year-old *P. bournei* seedlings in an open-top air chamber (OTC) under either an ambient CO_2 (aCO_2) concentration or an eCO_2 concentration, applied different amounts of nitrate and ammonium to the soil, and studied the effect of nitrogen form and concentration on gas exchange and chlorophyll fluorescence parameters, which are important indicators of photosynthetic performance [36,37]. Chlorophyll has the function of absorbing and transferring light energy, which is essential for photosynthesis [38].We explored the mechanism of nitrogen regulation from the aspects of photosynthetic chlorophyll content, which can reflect the photosynthetic capacity of plant leaves and has been shown to be significantly positively correlated with the net photosynthetic rate (Pn) [39].We also determined rubisco and rubisco activase (RCA) content because rubisco is the core enzyme in the Calvin cycle and RCA increases the proportion of active rubisco in photosynthetic cells [40]. The response of plants to eCO_2 is basically mediated by leaf photosynthesis, which is closely related to changes in leaf structure and chemical composition [41]. Furthermore, leaf anatomical structures best reflect the effects of environmental factors on plants and the adaptation strategies of plants to the environment [42]. Therefore, we also investigated the influence of the different treatments on leaf structure. These data should provide a theoretical basis and technical guidance for *P. bournei* cultivation in the future.

2. Materials and Methods

2.1. Plant Materials and Fertilizer

One-year-old *P. bournei* container seedlings grown from seeds of the same mother plant were used as the experimental material. In January 2019, seedlings (average height, 9.4 cm; average ground diameter, 2.73 mm) were selected and transplanted to pots (diameter 15.5 cm, height 14.0 cm) containing 1.5 kg of yellow soil, with one seedling per pot. The nutrient content of the soil was as follows: total nitrogen, 0.84g·kg^{-1}; alkali nitrogen, 28.3 mg·kg^{-1}; total phosphorus, 0.23 g·kg^{-1}; available phosphorus, 7.8 mg·kg^{-1}; total potassium, 21.6 g·kg^{-1}; available potassium, 207.3 mg·kg^{-1}. Seedlings were allowed to recover from transplanting for four months before treatments were applied. The nitrogen fertilizers used in the experiment were calcium nitrate and ammonium sulfate, the phosphate fertilizer was superphosphate, and the potassium fertilizer was potassium chloride.

2.2. Experimental Treatments

An OTC was used to simulate eCO_2 conditions. The OTC was a regular octagonal prism with sides of 1 m in length, a diameter of 2.42 m, and a height of 1.7 m, and comprised a 4 cm × 4 cm square steel frame and high-light-transmission 8-mm tempered glass. To accumulate CO_2 and slow the rate of CO_2 gas loss, the top opening of the OTC was inclined 45 degrees inward. The CO_2 concentration in the gas chamber was controlled by a CO_2 sensor device (GMP-220, Vaisala, Finland). The collected data were transmitted to a computer that controlled the switch of the solenoid valve of each gas chamber by running the programmed program to control the CO_2 concentration in the gas chamber within the target concentration range. Seedlings were treated with either a natural a$CO2$ concentration of 280–450 $\mu\text{mol·mol}^{-1}$ or a controlled eCO_2 concentration of 700 $\mu\text{mol·mol}^{-1}$, and with either a moderate level of nitrate (NO_3^-; 0.8 g per seedling), a high level of nitrate (iNO_3^-; 1.2 g per seedling), a moderate level of ammonium (NH_4^+; 0.8 g per seedling), or a high level of ammonium (iNH_4^+; 1.2 g per seedling), i.e., eight treatments in total, each treatment consisting of 45 replicates, totaling to 360 pots. The nitrogen applications were divided into three equal quantities and applied on 3 June, 3 July, and 2 August. The CO_2 treatments began on 12 June and endedon8 September. Phosphorus and potassium were also applied to seedlings on 3 June, 3 July, and 2 August. The phosphorus and potassium application rate was based on the moderate level of nitrogen application (i.e., N:P_2O_5:K_2O = 1:1:1). The relevant indicators for each treatment were assessed at the end of CO_2 treatment.

2.3. Determination of Gas Exchange

Three seedlings were selected from each treatment, and each of the three upperpart functional leaves were labeled. The net photosynthetic rate (Pn), stomatal conductance (Gs), intercellular CO_2 concentration (Ci), and transpiration rate (Tr) of these leaves were measured between 9:00 a.m. and 11:00 a.m. on a sunny day using an Li-6400 portable photosynthesis measurement system (LI-COR Biosciences, Lincoln, NE, USA). The LiCor chamber temperature and humidity measured by the LiCor were 30.7 ± 1.8 °C and 48.6 ± 2.4%. A small CO_2 cylinder was used to control the CO_2 concentration of the leaf chamber at 350 $\mu mol \cdot mol^{-1}$ in theaCO$_2$ gas chamber and at 700 $\mu mol \cdot mol^{-1}$ in the eCO$_2$chamber. Photon flux was set to 1200 $\mu mol \cdot mol^{-1}$ using a red and blue light source.

2.4. Determination of Chlorophyll Fluorescence

Chlorophyll fluorescence was determined using a modulated chlorophyll fluorescence imaging system (Imaging-Pam mini version, Walz, Germany). After seedlings had been treated in the dark for 1h, the initial fluorescence (Fo) and maximum fluorescence (Fm) in the dark-adapted state of the labeled leaves were determined. Next, the fluorescence kinetic curve was determined to calculate the electron transport rate (ETR), actual photochemical efficiency (Y(II)), photochemical quenching (qP), and non-photochemical quenching (NPQ) in which the qP is the quenching of chlorophyll fluorescence production due to photosynthesis and NPQ is the quenching of chlorophyll fluorescence production due to heat dissipation. The maximum quantum efficiency of photosystem II (Fv/Fm) was determined by calculating (Fm − Fo)/Fm. The Fo′ and Fm′ are initial fluorescence and maximum fluorescence in the light-adapted state respectively, the effective photochemical efficiency of photosystem II (Fv′/Fm′)was determined by calculating (Fm′ − Fo′)/Fm′.

2.5. Determination of Chlorophyll and Key Photosynthetic Enzyme content

The upper functional leaves (0.2 g) of five seedlings in each treatment were sampled. The chlorophyll content was determined by ethanol extraction UV spectrophotometry and calculated according to the following formula: chlorophyll $a = 12.7 \times OD_{663} - 2.69 \times OD_{645}$; chlorophyll $b = 22.9 \times OD_{645} - 4.68 \times OD_{663}$ [43], OD is optical density. Rubisco and RCA content were measured using enzyme-linked immunosorbent assay kits produced by Shanghai Keshun Biological Technology Co., Ltd., Shanghai, China.

2.6. Determination of Leaf Anatomy and Stomatal Density

To observe the leaf anatomy, labeled leaves were fixed in FAA (5% formaldehyde, 5% glacial acetic acid, 90% ethanol of 70%), embedded in paraffin wax and then sectioned to obtain transverse sections (10 μm thick), which were stained with safranin-fast green. These sections were dehydrated and their anatomy was observed using an optical microscope of Nikon Eclipse (Nikon, Tokyo, Japan) and photographed using a Nikon DS-U3 (Nikon, Tokyo, Japan) [44]. Case Viewer software was used to measure the following anatomical structure indicators: cell tightness rate (CTR) = palisade tissue thickness/leaf thickness; scattered rate (SR) = spongy tissue thickness/leaf thickness [45]. To calculate stomatal density, labeled fresh leaves were sprayed with gold and observed and photographed using a scanning electron microscope (Carl Zeiss, Jena, Germany).

2.7. Statistical Analysis

Three-way analysis of variance (ANOVA) was employed to compare the treatment effects, and the treatments means were evaluated by Tukey's multiple-range test ($p < 0.05$). All data analyses were performed using Statistical Product and Service Solutions (SPSS) version 23.0. Figures and tables were drafted using Microsoft Office 2007.

3. Results

3.1. Gas Exchange

Under eCO_2, the Pn, Gs, and Tr of seedlings treated with a moderate amount of nitrate were significantly lower (27.0%, 20.2%, and 22.5% lower, respectively) than that of seedlings grown under aCO_2 (Figure 1). However, compared with seedlings treated with a moderate amount of nitrate, seedlings treated with a high level of nitrate showed a significant increase in Pn, Gs, and Tr under both eCO_2 and aCO_2. Furthermore, the Pn, Gs, and Tr values obtained under aCO_2 and eCO_2 conditions were not significantly different (i.e., the adverse effects of eCO_2 on the gas exchange of *P. bournei* seedlings were alleviated when seedlings were treated with a high level of nitrate) (Figure 1). Seedlings that received applications of ammonium showed increased Pn, Gs, and Tr under eCO_2 compared with seedlings grown under aCO_2; however, the Pn of seedlings that received moderate and high applications of ammonium were only 8.3% and 4.4% higher, respectively, than those grown under aCO_2 and the difference was not significant. By contrast, the Gs of seedlings that received moderate and high applications of ammonium were 21.9% and 23.5% higher, respectively, under eCO_2 than those grown under aCO_2, indicating that under eCO_2, ammonium applications were beneficial for gas exchange (Figure 1).

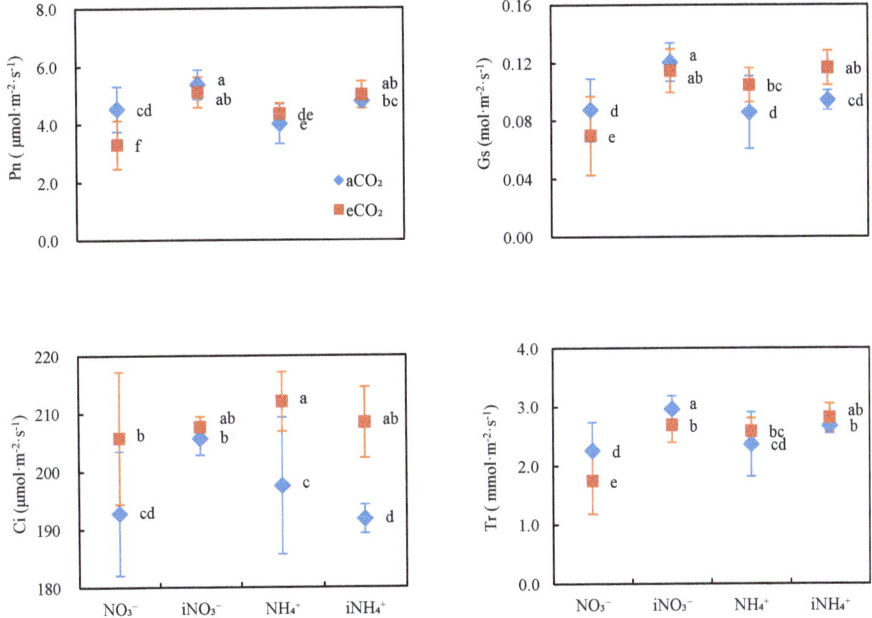

Figure 1. Gas exchange parameters of *Phoebe bournei* seedlings grown under ambient (aCO_2) or elevated CO_2 (eCO_2) concentrations and that received moderate (NO_3^-) or high (iNO_3^-) applications of nitrate or moderate (NH_4^+) or high (iNH_4^+) applications of ammonium. Pn, net photosynthetic rate; Gs, stomatal conductance; Ci, intercellular CO_2 concentration; Tr, transpiration rate. Data points represent means ± standard deviation (SD); $n = 9$. Different lowercase letters represent significant differences between treatments ($p < 0.05$) according to Tukey's test.

3.2. Chlorophyll Fluorescence

Under eCO_2, ETR and Y(II) of seedlings treated with moderate nitrate applications were significantly lower than under aCO_2; however, additional nitrate increased ETR and Y(II) regardless of CO_2 treatment (Figure 2). The ETR and Y(II) of seedlings applied with ammonium were higher under eCO_2 than under aCO_2. The qP values of seedlings treated with high nitrate applications were significantly higher than those treated with moderate nitrate or ammonium applications regardless of CO_2 treatment. By contrast, the qP of seedlings treated with high ammonium was only significantly higher (36.59% higher) than those treated with moderate ammonium or with moderate nitrate applications when grown under eCO_2 (Figure 2). The different CO_2 concentrations did not significantly affect the Fv/Fm, Fv′/Fm′, and NPQ values of seedlings (Figure 2).

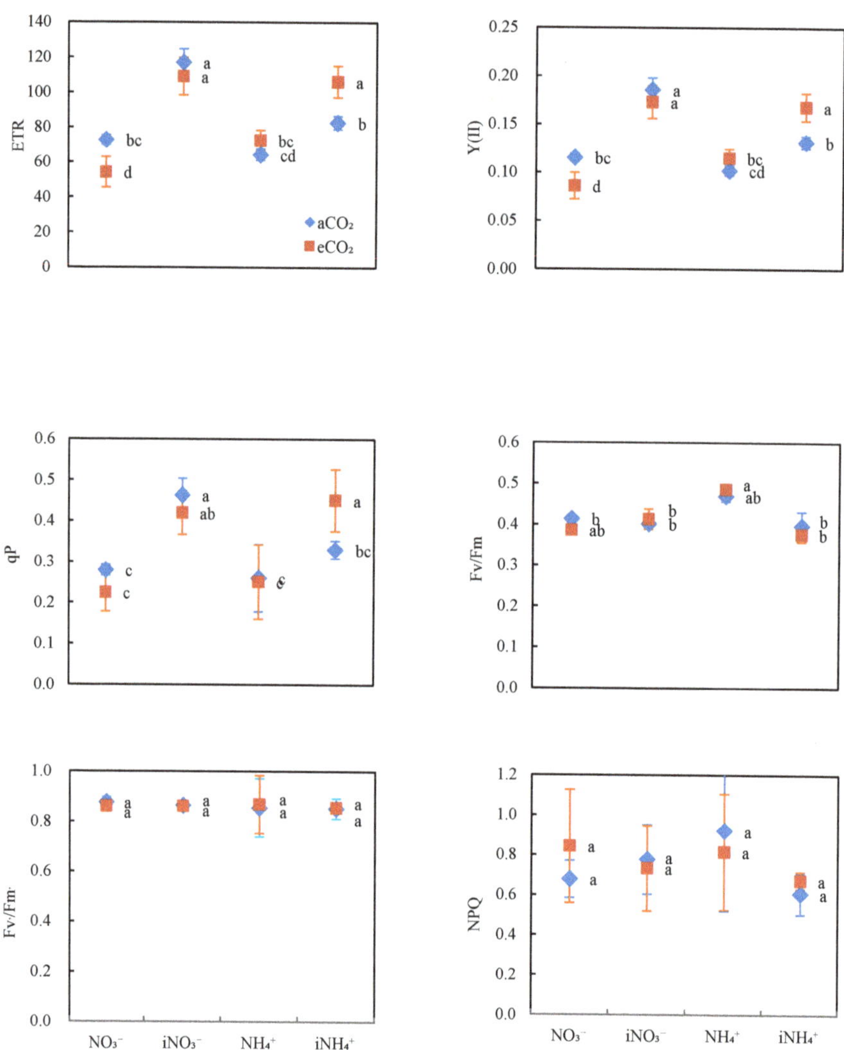

Figure 2. Fluorescence of Phoebe bournei seedlings grown under ambient (aCO_2) or elevated CO_2 (eCO_2) concentrations and that received moderate (NO_3^-) or high (iNO_3^-) applications of nitrate or moderate (NH_4^+) or high (iNH_4^+) applications of ammonium. ETR, electron transport rate; Y(II), actual photochemical efficiency; qP, photochemical quenching; Fv/Fm, maximum quantum efficiency of photosystem II; Fv′/Fm′, effective photochemical efficiency of photosystem II; NPQ, non-photochemical quenching. Data points represent means ± SD; n = 5. Data points with the same lowercase letter are not significantly different ($p > 0.05$) according to Tukey's test.

3.3. Chlorophyll Content

The chlorophyll content of seedlings treated with moderate nitrate applications decreased when grown under elevated CO_2 conditions (Table 1). The chlorophyll a, b, and $a + b$ content of seedlings that received moderate applications of nitrate decreased under eCO_2 by 19.3%, 20.0%, and 18.6%, respectively. By contrast, the chlorophyll a, b and $a + b$ content of seedlings grown under eCO_2 that received high applications of nitrate were not significantly different, indicating that high levels of nitrate could alleviate the negative effects of eCO_2 on chlorophyll content (Table 2). Seedlings treated with ammonium showed higher levels of chlorophyll a, b and $a + b$ synthesis under eCO_2 than seedlings grown under aCO_2, particularly when treated with high levels of ammonium. Although the chlorophyll a/b of seedlings was not significantly affected by the different CO_2 concentrations, chlorophyll a/b levels were significantly higher when treated with moderate nitrogen applications rather than high nitrogen applications.

Table 1. Effect of different CO_2 concentrations and nitrogen treatments on the chlorophyll a and b content of *Phoebe bournei* seedlings.

Treatment		Chl a (mg·g^{-1})	Chl b (mg·g^{-1})	Chl $a + b$ (mg·g^{-1})	Chl a/b
	NO_3^-	1.45 ± 0.06cd	0.50 ± 0.08de	1.94 ± 0.13d	2.95 ± 0.33a
aCO_2	iNO_3^-	1.67 ± 0.02a	0.83 ± 0.08a	2.50 ± 0.09a	2.02 ± 0.19d
	NH_4^+	1.38 ± 0.06d	0.48 ± 0.03de	1.86 ± 0.09d	2.91 ± 0.08a
	iNH_4^+	1.59 ± 0.05ab	0.63 ± 0.08cd	2.22 ± 0.12bc	2.53 ± 0.23abc
	NO_3^-	1.17 ± 0.09e	0.40 ± 0.06e	1.58 ± 0.15e	2.94 ± 0.30a
eCO_2	iNO_3^-	1.68 ± 0.03a	0.79 ± 0.08ab	2.47 ± 0.1ab	2.13 ± 0.18cd
	NH_4^+	1.50 ± 0.10bc	0.57 ± 0.09cd	2.08 ± 0.2cd	2.66 ± 0.28ab
	iNH_4^+	1.59 ± 0.08ab	0.67 ± 0.14bc	2.26 ± 0.21abc	2.42 ± 0.37bcd

Values represent means $\pm$ standard error(SE) of the mean; $n = 5$. Values followed by different letters within the same column are significantly different ($p < 0.05$) according to Tukey's test. Treatments: aCO_2, ambient CO_2; eCO_2, elevated CO_2; NO_3^-, moderate nitrate; iNO_3^-, high nitrate; NH_4^+, moderate ammonium; iNH_4^+, high ammonium.

Table 2. Analysis results of three factors (CO_2 concentration, nitrogen level, nitrogen type) of chlorophyll content for *Phoebe bournei* seedlings.

Effects	Chl a		Chl b		Chl $a + b$		Chl a/b	
	F	p	F	p	F	p	F	p
CO_2 concentration (C)	1.284	0.274	0.001	0.972	0.244	0.628	0.144	0.709
Nitrogen level (L)	77.804	<0.001	42.524	<0.001	62.589	<0.001	26.394	<0.001
Nitrogen Type (T)	0.570	0.461	1.220	0.286	0.100	0.755	1.087	0.313
C × L	2.746	0.117	0.061	0.809	0.812	0.381	0.093	0.764
C × T	10.238	0.006	2.801	0.114	6.033	0.026	0.781	0.390
L × T	10.753	0.005	7.648	0.014	9.891	0.006	4.861	0.042
C × L × T	9.126	0.008	0.271	0.610	2.842	0.111	0.003	0.955

F and p values of three-way ANOVA are given. CO_2 concentration, ambient CO_2, elevated CO_2; nitrogen level, nitrate, ammonium; nitrogen Type, moderate, high. F is the value of the F test statistic. $p < 0.01$ represents the difference was extremely significant, $p < 0.05$ represents the difference was significant.

3.4. Content of Key Photosynthetic Enzymes

The rubisco and RCA content of seedlings treated with a moderate amount of nitrate were significantly lower (21.48% and 38.81% lower, respectively) in seedlings grown under eCO_2 compared with those grown under aCO_2. By contrast, the rubisco and RCA content in seedlings treated with high levels of nitrate increased by 22.30% and 2.00%, respectively, under eCO_2 compared with seedlings under aCO_2 (Figure 3). The rubisco and RCA content of seedlings that received high levels of ammonium were significantly higher than those that received moderate ammonium application, but were not significantly different under aCO_2 or eCO_2 conditions (Figure 3).

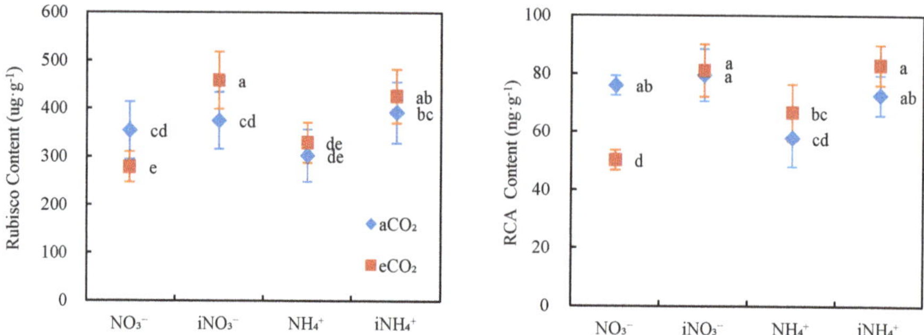

Figure 3. Rubisco and rubisco activase (RCA) content of *Phoebe bournei* seedlings grown under ambient (aCO$_2$) or elevated CO$_2$ (eCO$_2$) concentrations and that received moderate (NO$_3^-$) or high (iNO$_3^-$) applications of nitrate or moderate (NH$_4^+$) or high (iNH$_4^+$) applications of ammonium. Data points represent means ± SE; $n = 5$. Different letters beside data points represent significant differences between treatments ($p < 0.05$) according to Tukey's test.

3.5. Leaf Anatomical Structure and Stomatal Density

We assessed the upper and lower epidermis, palisade tissue, spongy tissue structure and stomata (Figure 4). The leaf thickness and stomatal density values of seedlings treated with moderate nitrate applications and grown under eCO$_2$ were significantly lower (12.44% and 18.90% lower, respectively) than those of seedlings grown under aCO$_2$ (Table 3). High nitrate applications had a positive effect on stomatal density; however, high nitrate applications did not have a significant effect on leaf thickness under eCO$_2$. The leaf thickness of seedlings that were treated with ammonium and grown under eCO$_2$ was not significantly different to those grown under aCO$_2$ (Table 3). Moderate application of ammonium did not significantly affect stomatal density; however, high levels of ammonium significantly increased stomatal density under eCO$_2$ compared with seedlings grown under aCO$_2$ (Table 3). The different nitrogen and CO$_2$ treatments did not significantly affect palisade tissue thickness and CTR.

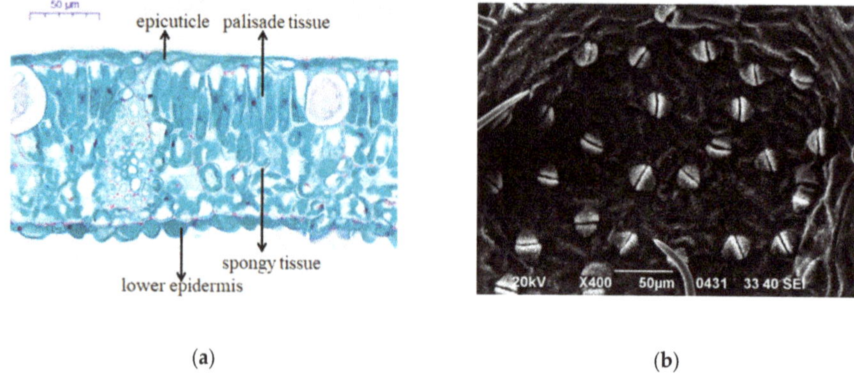

(**a**)   (**b**)

Figure 4. Examples of leaf anatomical structure and stomata of *Phoebe bournei* seedlings grown under ambient (aCO$_2$) or elevated CO$_2$ (eCO$_2$) concentrations and that received moderate (NO$_3^-$) or high (iNO$_3^-$) applications of nitrate or moderate (NH$_4^+$) or high (iNH$_4^+$) applications of ammonium. (**a**) Leaf anatomical structure; (**b**) leaf stomata.

Table 3. Effect of different treatments on the leaf anatomical structure and stomatal density of *Phoebe bournei* seedlings.

Treatment		Palisade Tissue Thickness (µm)	Spongy Tissue Thickness (µm)	Epidermal Thickness (µm)	Leaf Thickness (µm)	CTR	SR	Stomatal Density (ind·mm^{-2})
aCO$_2$	NO$_3^-$	38.8 ± 4.9a	51.7 ± 5.7ab	11.2 ± 0.9b	108.5 ± 2.8bc	0.36 ± 0.04a	378.8 ± 12.6d	378.8 ± 12.6d
	iNO$_3^-$	41.1 ± 3.9a	58.5 ± 5.1a	11.7 ± 0.5b	113.8 ± 4ab	0.36 ± 0.04a	446.1 ± 14.6c	446.1 ± 14.6c
	NH$_4^+$	37.2 ± 3.4a	49.4 ± 4.7ab	11.5 ± 0.7b	100.1 ± 3.8d	0.37 ± 0.05a	383.0 ± 26.3d	383.0 ± 26.3d
	iNH$_4^+$	39.5 ± 4.1a	48.4 ± 4ab	13.9 ± 1.5ab	101.2 ± 0.8cd	0.39 ± 0.04a	399.8 ± 14.6d	399.8 ± 14.6d
eCO$_2$	NO$_3^-$	37.0 ± 6.4a	42.4 ± 4.3b	11.7 ± 1.1b	95.0 ± 6.9d	0.39 ± 0.05a	307.2 ± 7.3e	307.2 ± 7.3e
	iNO$_3^-$	35.5 ± 3.2a	56.2 ± 3.6a	16.0 ± 0.8a	115.9 ± 6.8a	0.31 ± 0.02a	542.9 ± 25.2a	542.9 ± 25.2a
	NH$_4^+$	37.3 ± 6.9a	53.2 ± 6.6a	12.6 ± 1.5b	101.7 ± 0.7cd	0.37 ± 0.07a	374.6 ± 26.3d	374.6 ± 26.3d
	iNH$_4^+$	40.9 ± 1.1a	52.4 ± 1.8ab	12.4 ± 1.7b	108.1 ± 0.8bc	0.38 ± 0.01a	492.4 ± 25.3b	492.4 ± 25.3b

Treatments: aCO$_2$, ambient CO$_2$; eCO$_2$, elevated CO$_2$; NO$_3^-$, moderate nitrate; iNO$_3^-$, high nitrate; NH$_4^+$, moderate ammonium; iNH$_4^+$, high ammonium. CTR, cell tightness rate; SR, scattered rate. Values represent means ± SE of mean; $n = 3$. Values followed by different letters within the same column are significantly different ($p < 0.05$) according to Tukey's test.

4. Discussion

Although the photosynthetic rate of various plants grown under eCO$_2$ has been shown to increase in the short-term, generally, under long-term eCO$_2$, the stimulation effect weakens but the rates under eCO$_2$ remain higher than controls [19]. The growth response of trees to eCO$_2$ manifests in many physiological aspects in addition to photosynthetic efficiency. Although long-term eCO$_2$ has been shown to decrease the Gs of *Liquidambar styraciflua*, stomatal density was unaffected [46]. However, a survey of 100 species and 122 observations showed an average reduction in stomatal density of 14.3% with CO$_2$ enrichment [47]. Elevated CO$_2$ and limited nitrogen nutrition can restrict excitation energy dissipation in photosystem II of *Betula platyphylla* var. *japonica* [48] and expression of rubisco and RCA in *Pinus ponderosa* [49] and *Piceaabies* [50]. In this study, the photosynthetic efficiency of seedlings treated with a moderate amount of nitrate decreased significantly under eCO$_2$. In addition, the Gs, ETR, chlorophyll content, rubisco and RCA content, and leaf structure-related indexes of seedlings were reduced, indicating that under eCO$_2$ and moderate application of nitrate, the anatomical structure and photosynthetic physiological state of seedlings changed, causing photosynthesis to decrease. However, we found that there was no CO$_2$ effect under high applications of nitrate. By contrast, the Gs, ETR, Y (II), qP, and chlorophyll content did not change significantly, and rubisco and RCA content and stomatal density increased significantly, indicating that high nitrate applications under eCO$_2$ could help to improve the photosynthetic physiological and structural characteristics of *P. bournei* seedlings to sustain stimulation of photosynthesis.

Plant growth is governed by the balance of carbon assimilation and respiration. The joint influence of low nitrogen and excess light had an adverse effect on plant growth; in contrast, no adverse physiological responses were observed for plants under either nitrogen limitation or high light intensity conditions [25]. Similarly, in this study, *P. bournei* seedlings that received moderate nitrate applications under eCO$_2$ showed the phenomenon of photosynthetic decline, which is consistent with the findings reported by Han et al. [35]. High nitrate applications improved the physiological and structural characteristics of leaves of *P. bournei* seedlings and reduced photosynthetic decline. A possible explanation for this is that adequate nitrogen supply involved in carbohydrate catabolism and ATP production, which was in line with the sugar availability and high photosynthetic activity and so facilitated plants to cope with adverse environments such as abiotic stress. High applications of nitrate compensated for the shortage of an external nitrogen supply to some extent and adjusted the state of the carbon and nitrogen supply balance, which resulted in improved leaf structure and physiological characteristics and alleviated the degree of photosynthesis decline. Nitrogen is the basic element for enzyme synthesis in plants. About 40% of available nitrogen might be in rubisco [51]. Changes in nitrogen supply had a significant effect on rubisco and RCA content levels, and ultimately on the photosynthetic decline of *P. bournei* seedlings. Another possible explanation for the photosynthetic decline of *P. bournei* seedlings under eCO$_2$ is that eCO$_2$ affects the metabolism and absorption of nitrate in plants [52,53], which results in the photosynthetic decline of *P. bournei* seedlings, whereas higher applications of nitrate alleviated the negative impact of eCO_2 on nitrate metabolism and absorption to a certain extent so as to slow down photosynthetic decline. A study by Pettersson and McDonald [26] showed that an insufficient supply of nitrogen could reduce the expression of enzymes involved in photosynthesis, and limit the synthesis of proteins, leading to photosynthetic decline. Furthermore, eCO_2 concentrations can have certain inhibitory effects on the plant's photosynthetic rate, chlorophyll content, photosynthetic electron transport and potential activity of PSII, and its negative effects can be effectively alleviated by nitrogen application [54].

One involves the first biochemical step of nitrate assimilation, the conversion of nitrate to nitrite in the cytoplasm of leaf mesophyll cells. Photorespiration stimulates the export of malate from chloroplasts and increases the availability of nicotinamide adenine dinucleotide hydride in the cytoplasm that powers this first step of nitrate assimilation [22]. The eCO_2 conditions can reduce the photo respiration of leaves [55] and limit plant nitrate uptake in favor of ammonium uptake because eCO_2 conditions inhibit plant photorespiration and thus affect the transport of malate, which is ultimately detrimental to the assimilation of nitrate [56,57]. Under eCO_2, the Pn of seedlings treated with ammonium was increased compared with those grown under aCO_2, photosynthetic decline did not occur, and leaf physiological and structural indicators (including ETR, Y(II), and the content of chlorophyll and key photosynthetic enzymes) were increased, which demonstrated that the application of ammonium under eCO_2 was more beneficial for the growth of *P. bournei* seedlings compared with aCO_2. However, leaf physiological and structural indicators of seedlings treated with a moderate amount of nitrate decreased significantly under eCO_2; the reason for this finding may be that eCO_2 affects photorespiration, which in turn affects the assimilation of nitrate but has no effect on the assimilation of ammonium. Silva et al. [58] reported that coffee trees grown under eCO_2 and treated with ammonium showed better growth than those treated with nitrate. Some studies [32,33,59] have also indicated that ammonium treatment was superior to nitrate treatment under eCO_2 conditions because the increase in CO_2 concentration affected plant stomata conductance and water transpiration, which in turn affected the movement of nitrate in the soil with water, and ultimately affected the absorption of nitrate by the root system. This may also be the reason that ammonium had a more positive effect on seedlings than nitrate under eCO_2 in our experiment. High applications of ammonium had a better effect on *P. bournei* seedlings than moderate applications. The reason for this may be that an unsaturated increase in fertilizer supply promoted growth. Under the same CO_2 concentration, the photosynthesis of *P. bournei* seedlings that received an ammonium treatment was weaker than that of seedlings that were treated with nitrate, which may be due to a preference of *P. bournei* seedlings for this form of nitrogen. Although nitrogen limitation is a major bottleneck for many terrestrial ecosystems, phosphorus limitation issues are also critical for several ecosystems and species [60]. Phosphorus affects plant responses to eCO_2 [61,62]. Also, eCO_2 results in lower Zn and Fe assimilation rates [63]. Therefore, further studies that consider other elements could be recommended in the future.

5. Conclusions

The results of this experiment indicate that the photosynthetic efficiency of *P. bournei* seedlings treated with a moderate amount of nitrate decreased under eCO_2 conditions and that photosynthetic decline occurred. Increasing the dosage of nitrate or the supply of ammonium fertilizer could regulate the physiological and structural characteristics of *P. bournei* seedlings and alleviate or avoid photosynthetic decline, which would be beneficial for the photosynthesis of *P. bournei* seedlings. Therefore, we suggest that increasing the application of nitrate or applying ammonium may be an effective way of improving the growth of *P. bournei* seedlings under eCO_2 levels in the future.

Author Contributions: X.W. (Xiao Wang) and X.W. (Xiaoli Wei) conceived this study; X.W. (Xiao Wang) and S.C. carried out the experiments; X.W. (Xiao Wang), X.W. (Xiaoli Wei) and G.W. analyzed the results; X.W. (Xiao Wang) wrote the manuscript with input from the other authors. All authors have read and agreed to the published version of the manuscript.

Funding: This research was funded by the Guizhou Province Technology Plan Project, QianKe He Foundation (2019)1407 and Guizhou Province High-level Innovative Talents Training Plan Project (2016) 5661.

Acknowledgments: We thank Chi Wu for assistance with conducting the experiments.

Conflicts of Interest: The authors declare no conflict of interest.

References

1. Landry, J.S.; Matthews, H.D. Non-deforestation fire vs. fossil fuel combustion: The source of CO_2 emissions affects the global carbon cycle and climate responses. *Biogeosciences* **2016**, *13*, 2137–2149. [CrossRef]

2. De Graaff, M.A.; Van Groenigen, K.J.; Six, J.; Hungate, B.; Van Kessel, C. Interactions between plant growth and soil nutrient cycling under elevated CO_2: A meta-analysis. *Glob. Chang. Biol.* **2006**, *12*, 2077–2091. [CrossRef]

3. Aranjuelo, I.; Pardo, A.; Biel, C.; Savé, R.; Azcón-bieto, J.; Nogués, S. Leaf carbon management in slow-growing plants exposed to elevated CO_2. *Glob. Chang. Biol.* **2009**, *15*, 97–109. [CrossRef]

4. Intergovernmental Panel on Climate Change. *Climate Change 2014: Impacts, Adaptation, and Vulnerability. Part B: Regional Aspects*; Contribution of Working Group II to the Fifth Assessment Report of the Intergovernmental Panel on Climate Change; Cambridge University Press: Cambridge, UK, 2014.

5. Norby, R.J.; Zak, D.R. Ecological lessons from free-air CO_2 enrichment (FACE) experiments. *Annu. Rev. Ecol. Evol. Syst.* **2011**, *42*, 181–203. [CrossRef]

6. Long, S.P.; Ainsworth, E.A.; Rogers, A.; Ort, D.R. Rising atmospheric carbon dioxide: Plants FACE the future. *Annu. Rev. Plant Biol.* **2004**, *55*, 591–628. [CrossRef] [PubMed]

7. Zhu, P.; Zhuang, Q.; Ciais, P.; Welp, L.; Li, W.; Xin, Q. Elevated atmospheric CO_2 negatively impacts photosynthesis through radiative forcing and physiology-mediated climate feedback. *Geophys. Res. Lett.* **2017**, *44*, 1956–1963. [CrossRef]

8. Yu, T.; Chen, Y. Effects of elevated carbon dioxide on environmental microbes and its mechanisms: A review. *Sci. Total Environ.* **2019**, *655*, 865–879. [CrossRef]

9. Quentin, A.G.; Crous, K.Y.; Barton, C.V.; Ellsworth, D.S. Photosynthetic enhancement by elevated CO_2 depends on seasonal temperatures for warmed and non-warmed *Eucalyptus globulus* trees. *Tree Physiol.* **2015**, *35*, 1249–1263.

10. DaMatta, F.M.; Godoy, A.G.; Menezes-Silva, P.E.; Martins, S.C.; Sanglard, L.M.; Morais, L.E.; Torre-Neto, A.; Ghini, R. Sustained enhancement of photosynthesis in coffee trees grown under free-air CO_2 enrichment conditions: Disentangling the contributions of stomatal, mesophyll, and biochemical limitations. *J. Exp. Bot.* **2016**, *67*, 341–352. [CrossRef]

11. Davidson, A.M.; Da Silva, D.; Saa, S.; Mann, P.; DeJong, T.M. The influence of elevated CO_2 on the photosynthesis, carbohydrate status, and plastochron of young peach (*Prunuspersica*) trees. *Hortic. Environ. Biotechnol.* **2016**, *57*, 364–370. [CrossRef]

12. Dosskey, M.G.; Linderman, R.G.; Boersma, L. Carbon–sink stimulation of photosynthesis in Douglas fir seedlings by some ectomycorrhizas. *New Phytol.* **1990**, *115*, 269–274. [CrossRef]

13. Cure, J.D.; Acock, B. Crop responses to carbon dioxide doubling: A literature survey. *Agric. For. Meteorol.* **1986**, *38*, 127–145. [CrossRef]

14. Harley, P.C.; Thomas, R.B.; Reynolds, J.F.; Strain, B.R. Modelling photosynthesis of cotton grown in elevated CO_2. *Plant Cell Environ.* **2010**, *15*, 271–282. [CrossRef]

15. Cheng, S.H.; d Moore, B.; Seemann, J.R. Effects of short-and long-term elevated CO_2 on the expression of ribulose-1, 5-bisphosphate carboxylase/oxygenase genes and carbohydrate accumulation in leaves of *Arabidopsis thaliana* (L.) Heynh. *Plant Physiol.* **1998**, *116*, 715–723. [CrossRef]

16. Vicente, R.; Pérez, P.; Martínez-Carrasco, R.; Usadel, B.; Kostadinova, S.; Morcuende, R. Quantitative RT-PCR platform to measure transcript levels of C and N metabolism-related genes in durum wheat: Transcript profiles in elevated CO_2 and high temperature at different levels of N supply. *Plant Cell Physiol.* **2015**, *56*, 1556–1573. [CrossRef] [PubMed]

17. Roumet, C.; Garnier, E.; Hélène, S.; Salager, J.L.; Roy, J. Short and long-term responses of whole-plant gas exchange to elevated CO_2 in four herbaceous species. *Environ. Exp. Bot.* **2000**, *43*, 155–169. [CrossRef]

18. Ainsworth, E.A.; Davey, P.A.; Hymus, G.J.; Osborne, C.P.; Long, S.P. Is stimulation of leaf photosynthesis by elevated carbon dioxide concentration maintained in the long term? A test with *Loliumperenne* grown for 10 years at two nitrogen fertilization levels under Free Air CO_2 Enrichment (FACE). *Plant Cell Environ.* **2003**, *26*, 705–714. [CrossRef]

19. Seneweera, S. Effects of elevated CO_2 on plant growth and nutrient partitioning of rice (*Oryza sativa* L.) at rapid tillering and physiological maturity. *J. Plant Interact.* **2011**, *6*, 35–42. [CrossRef]

20. Alonso, A.; Pérez, P.; Martínez-Carrasco, R. Growth in elevated CO_2 enhances temperature response of photosynthesis in wheat. *Physiol. Plant.* **2009**, *135*, 109–120. [CrossRef]

21. Gutiérrez, D.; Gutiérrez, E.; Pérez, P.; Morcuende, R.; Verdejo, A.L.; Martinez-Carrasco, R. Acclimation to future atmospheric CO_2 levels increases photochemical efficiency and mitigates photochemistry inhibition by warm temperatures in wheat under field chambers. *Physiol. Plant.* **2009**, *137*, 86–100. [CrossRef]

22. Bloom, A.J.; Asensio, J.S.R.; Randall, L.; Rachmilevitch, S.; Cousins, A.B.; Carlisle, E.A. CO_2 enrichment inhibits shoot nitrate assimilation in C3 but not C4 plants and slows growth under nitrate in C3 plants. *Ecology* **2012**, *93*, 355–367. [CrossRef] [PubMed]

23. Harbinson, J.; Genty, B.; Baker, N.R. The relationship between CO_2 assimilation and electron transport in leaves. *Photosynth. Res.* **1990**, *25*, 213–224. [CrossRef] [PubMed]

24. Yousuf, P.Y.; Ganie, A.H.; Khan, I.; Qureshi, M.I.; Ibrahim, M.M.; Sarwat, M.; Iqbal, M.; Ahmad, A. Nitrogen-efficient and nitrogen-inefficient Indian mustard showed differential expression pattern of proteins in response to elevated CO_2 and low nitrogen. *Front. Plant Sci.* **2016**, *7*, 1074. [CrossRef] [PubMed]

25. Cohen, I.; Rapaport, T.; Chalifa-Caspi, V.; Rachmilevitch, S. Synergistic effects of abiotic stresses in plants: A case study of nitrogen limitation and saturating light intensity in Arabidopsis thaliana. *Physiol. Plant.* **2019**, *165*, 755–767. [CrossRef]

26. Pettersson, R.; McDonald, A.J.S. Effects of nitrogen supply on the acclimation of photosynthesis to elevated CO_2. *Photosynth. Res.* **1994**, *39*, 389–400. [CrossRef]

27. Norby, R.J.; Warren, J.M.; Iversen, C.M.; Medlyn, B.E.; Mcmurtrie, R.E. CO_2 enhancement of forest productivity constrained by limited nitrogen availability. *Proc. Natl. Acad. Sci. USA* **2010**, *107*, 19368–19373. [CrossRef]

28. Jifon, J.L.; Wolfe, D.W. Photosynthetic acclimation to elevated CO_2 in *Phaseolus vulgaris* L. is altered by growth response to nitrogen supply. *Glob. Chang. Biol.* **2002**, *8*, 1018–1027. [CrossRef]

29. Makino, A. Rubisco and nitrogen relationships in rice: Leaf photosynthesis and plant growth. *Soil Sci. Plant Nutr.* **2003**, *49*, 319–327. [CrossRef]

30. Butterly, C.R.; Armstrong, R.; Chen, D.; Tang, C. Carbon and nitrogen partitioning of wheat and field pea grown with two nitrogen levels under elevated CO_2. *Plant Soil* **2015**, *391*, 367–382. [CrossRef]

31. Wujeska-Klause, A.; Crous, K.Y.; Ghannoum, O.; Ellsworth, D.S. Lower photorespiration in elevated CO_2 reduces leaf N concentrations in mature Eucalyptus trees in the field. *Glob. Chang. Biol.* **2019**, *25*, 1282–1295. [CrossRef]

32. Hachiya, T.; Sugiura, D.; Kojima, M.; Sato, S.; Yanagisawa, S.; Sakakibara, H.; Terashima, I.; Noguchi, K. High CO_2 triggers preferential root growth of *Arabidopsis thaliana* via two distinct systems under low pH and low N stresses. *Plant Cell Physiol.* **2014**, *55*, 269–280. [CrossRef] [PubMed]

33. Jauregui, I.; Aroca, R.; Garnica, M.; Zamarreño, Á.M.; García-Mina, J.M.; Serret, M.D.; Parry, M.; Iigoyen, J.J.; Aranjuelo, I. Nitrogen assimilation and transpiration: Key processes conditioning responsiveness of wheat to elevated CO_2 and temperature. *Physiol. Plant.* **2015**, *155*, 338–354. [CrossRef] [PubMed]

34. Andrews, M.; Condron, L.M.; Kemp, P.D.; Topping, J.F.; Lindsey, K.; Hodge, S.; Raven, J.A. Elevated CO_2 effects on nitrogen assimilation and growth of C3 vascular plants are similar regardless of N-form assimilated. *J. Exp. Bot.* **2018**, *70*, 683–690. [CrossRef] [PubMed]

35. Han, W.J.; Liao, F.Y.; He, P. The photosynthetic response of *Phoebe bournei* to doubled CO_2 concentration in air. *J. Cent. South Univ. For. Tech.* **2003**, *23*, 62–65.

36. Laisk, A.; Loreto, F. Determining photosynthetic parameters from leaf CO_2 exchange and chlorophyll fluorescence (ribulose-1, 5-bisphosphate carboxylase/oxygenase specificity factor, dark respiration in the light, excitation distribution between photosystems, alternative electron transport rate, and mesophyll diffusion resistance. *Plant Physiol.* **1996**, *110*, 903–912.

37. Peterson, R.B.; Oja, V.; Laisk, A. Chlorophyll fluorescence at 680 and 730 nm and leaf photosynthesis. *Photosynth. Res.* **2001**, *70*, 185–196. [CrossRef]

38. Frank, H.A.; Cua, A.; Chynwat, V.; Young, A.; Gosztola, D.; Wasielewski, M.R. Photophysics of the carotenoids associated with the xanthophyll cycle in photosynthesis. *Photosynth. Res.* **1994**, *41*, 389–395. [CrossRef]

39. Zhao, X.; Du, Q.; Zhao, Y.; Wang, H.; Yu, H. Effects of different potassium stress on leaf photosynthesis and chlorophyll fluorescence in maize (*Zea mays* L.) at seedling stage. *Agric. Sci.* **2016**, *7*, 44–53.

40. Zheng, B.; Cheng, X.; Jiang, D.; Weng, X. Effects of potassium on Rubisco, RCA and photosynthetic rate of plant. *J. Zhejiang For. Coll.* **2002**, *19*, 104–108.

41. Zheng, Y.; Li, F.; Hao, L.; Yu, J.; Xu, M. Elevated CO_2 concentration induces photosynthetic down-regulation with changes in leaf structure, non-structural carbohydrates and nitrogen content of soybean. *BMC Plant Biol.* **2019**, *19*, 255. [CrossRef]

42. Tomás, M.; Jaume, F.; Lucian, C.; Galmés, J.; Lea, H.; Medrano, H. Importance of leaf anatomy in determining mesophyll diffusion conductance to CO_2 across species: Quantitative limitations and scaling up by models. *J. Exp. Bot.* **2013**, *64*, 2269–2281. [CrossRef] [PubMed]

43. Yang, G.; Luo, J.; Lin, Y.; Wang, H.; Mo, Y. Effects of salt stress on photosynthetic characteristics and stoma structure of two kinds bracket plants. *Hubei Agric. Sci.* **2016**, *55*, 4982–4986.

44. Li, X.; Yang, Y.; Zhang, J.; Jia, L.; Li, Q.; Zhang, T.; Qiao, K.; Ma, S. Zinc induced phytotoxicity mechanism involved in root growth of *Triticum aestivum* L. *Ecotoxicol. Environ. Saf.* **2012**, *86*, 198–203. [CrossRef] [PubMed]

45. Wang, S.; Zou, Y.; Ma, F. Influence of drought stress on leaf anatomical structure and micro-morphology traits and chloroplast ultra structure of three *Malus* specie. *Agric. Res. Arid Areas* **2014**, *32*, 15–23.

46. Herrick, J.D.; Maherali, H.; Thomas, R.B. Reduced stomatal conductance in sweetgum (*Liquidambar styraciflua*) sustained over long-term CO_2 enrichment. *New Phytol.* **2004**, *162*, 387–396. [CrossRef]

47. Woodward, F.I.; Kelly, C.K. The influence of CO_2 concentration on stomatal density. *New Phytol.* **1995**, *131*, 311–327. [CrossRef]

48. Kitao, M.; Koike, T.; Tobita, H.; Maruyama, Y. Elevated CO_2 and limited nitrogen nutrition can restrict excitation energy dissipation in photosystem II of Japanese white birch (*Betula platyphylla* var. *japonica*) leaves. *Physiol. Plant.* **2005**, *125*, 64–73. [CrossRef]

49. Tissue, D.T.; Griffin, K.L.; Ball, J.T. Photosynthetic adjustment in field-grown ponderosa pine trees after six years of exposure to elevated CO_2. *Tree Physiol.* **1999**, *19*, 221–228. [CrossRef]

50. Urban, O.; Hrstka, M.; Zitová, M.; Holišová, P.; Šprtová, M.; Klem, K.; Calfapietra, C.; Angelis, P.D.; Marek, M.V. Effect of season, needle age and elevated CO_2 concentration on photosynthesis and Rubisco acclimation in *Piceaabies*. *Plant Physiol. Biochem.* **2012**, *58*, 135–141. [CrossRef]

51. Warren, C.R.; Adams, M.A. Distribution of N, Rubisco and photosynthesis in *Pinus pinaster* and acclimation to light. *Plant Cell Environ.* **2001**, *24*, 597–609. [CrossRef]

52. Causin, H.F.; Tremmel, D.C.; Rufty, T.W.; Reynolds, J.F. Growth, nitrogen uptake, and metabolism in two semiarid shrubs grown at ambient and elevated atmospheric CO_2 concentrations: Effects of nitrogen supply and source. *Am. J. Bot.* **2004**, *91*, 565–572. [CrossRef] [PubMed]

53. Alexandre, A.; Silva, J.; Buapet, P.; Björk, M.; Santos, R. Effects of CO_2 enrichment on photosynthesis, growth, and nitrogen metabolism of the sea grass *Zosteranoltii*. *Ecol. Evol.* **2012**, *2*, 2625–2635. [CrossRef] [PubMed]

54. Sanz-Sáez, Á.; Erice, G.; Aranjuelo, I.; Nogués, S.; Irigoyen, J.J.; Sánchez-Díaz, M. Photosynthetic down-regulation under elevated CO_2 exposure can be prevented by nitrogen supply in nodulated alfalfa. *J. Plant Physiol.* **2010**, *167*, 1558–1565. [CrossRef] [PubMed]

55. Sharkey, T.D. Estimating the rate of photorespiration in leaves. *Physiol. Plant.* **1988**, *73*, 147–152. [CrossRef]

56. Rachmilevitch, S.; Cousins, A.B.; Bloom, A.J. Nitrate assimilation in plant shoots depends on photorespiration. *Proc. Natl. Acad. Sci. USA* **2004**, *101*, 11506–11510. [CrossRef]

57. Bloom, A.J.; Burger, M.; Kimball, B.A.; Pinter, P.J., Jr. Nitrate assimilation is inhibited by elevated CO_2 in field-grown wheat. *Nat. Clim. Chang.* **2014**, *4*, 477–480. [CrossRef]

58. Silva, L.C.; Salamanca-Jimenez, A.; Doane, T.A.; Horwath, W.R. Carbon dioxide level and form of soil nitrogen regulate assimilation of atmospheric ammonia in young trees. *Sci. Rep.* **2015**, *5*, 13141. [CrossRef]

59. McGrath, J.M.; Lobell, D.B. Reduction of transpiration and altered nutrient allocation contribute to nutrient decline of crops grown in elevated CO_2 concentrations. *Plant Cell Environ.* **2013**, *36*, 697–705. [CrossRef]

60. Wassen, M.J.; Venterink, H.O.; Lapshina, E.D.; Tanneberger, F. Endangered plants persist under phosphorus limitation. *Nature* **2005**, *437*, 547–550. [CrossRef]

61. Conroy, J.P.; Milham, P.J.; Reed, M.L.; Barlow, E.W. Increases in phosphorus requirements for CO_2-enriched pine species. *Plant Physiol.* **1990**, *92*, 977–982. [CrossRef]

62. Israel, D.W.; Rufty, T.W., Jr.; Cure, J.D. Nitrogen and phosphorus nutritional interactions in a CO_2 enriched environment. *J. Plant Nutr.* **1990**, *13*, 1419–1433. [CrossRef]

63. Myers, S.S.; Zanobetti, A.; Kloog, I.; Huybers, P.; Leakey, A.D.; Bloom, A.J.; Carlisle, E.; Dietterich, L.H.; Fitzgerald, G.; Hasegawa, T.; et al. Increasing CO_2 threatens human nutrition. *Nature* **2014**, *510*, 139–142. [CrossRef] [PubMed]

Review

Adaptation of Forest Trees to Rapidly Changing Climate

Joanna Kijowska-Oberc [1,*], Aleksandra M. Staszak [2], Jan Kamiński [1] and Ewelina Ratajczak [1,*]

[1] Institute of Dendrology, Polish Academy of Sciences, Parkowa 5, 62-035 Kornik, Poland; jan_kaminski616@onet.pl

[2] Laboratory of Plant Physiology, Department of Plant Biology and Ecology, Faculty of Biology, University of Bialystok, Ciolkowskiego 1J, 15-245 Bialystok, Poland; a.staszak@uwb.edu.pl

* Correspondence: joberc@man.poznan.pl (J.K.-O.); eratajcz@man.poznan.pl (E.R.)

Received: 25 November 2019; Accepted: 19 January 2020; Published: 21 January 2020

Abstract: Climate change leads to global drought-induced stress and increased plant mortality. Tree species living in rapidly changing climate conditions are exposed to danger and must adapt to new climate conditions to survive. Trees respond to changes in the environment in numerous ways. Physiological modulation at the seed stage, germination strategy and further development are influenced by many different factors. We review forest abiotic threats (such as drought and heat), including biochemical responses of plants to stress, and biotic threats (pathogens and insects) related to global warming. We then discus the varied adaptations of tree species to changing climate conditions such as seed resistance to environmental stress, improved by an increase in temperature, affinity to specific fungal symbionts, a wide range of tolerance to abiotic environmental conditions in the offspring of populations occurring in continental climate, and germination strategies closely linked to the ecological niche of the species. The existing studies do not clearly indicate whether tree adaptations are shaped by epigenetics or phenology and do not define the role of phenotypic plasticity in tree development. We have created a juxtaposition of literature that is useful in identifying the factors that play key roles in these processes. We compare scientific evidence that species distribution and survival are possible due to phenotypic plasticity and thermal memory with studies that testify that trees' phenology depends on phylogenesis, but this issue is still open. It is possible that studies in the near future will bring us closer to understanding the mechanisms through which trees adapt to stressful conditions, especially in the context of epigenetic memory in long-lived organisms, and allow us to minimize the harmful effects of climatic events by predicting tree species' responses or by developing solutions such as assisted migration to mitigate the consequences of these phenomena.

Keywords: global climate change; forest ecology; trees adaptation; phenotypic plasticity

---

1. Introduction

Recent changes in the global climate have a considerable impact on many aspects of the natural environment and various aspects of human economy [1,2]. In addition, there is a fierce social and scientific debate concerning the origin of the climate changes that are presently being observed [3]. The results of many scientific studies show that human activity has contributed to the increase in CO_2 in the atmosphere; increased CO_2 levels cause an increase in the frequency of extreme weather events [4], and such events were not seen as often before 2010 as they have been since then. For example, between May and July 2018, almost every day, approximately 22% of the populated and agricultural areas north of 30° latitude experienced high temperature extremes at the same time [5]. Over a wide range of climate changes, the most severe are those related to changes in the temperature range. Forecasts for the coming decades predict that approximately a third of the permafrost will disappear after the global surface temperature rises by 2 °C [6]. On the one hand, water levels are increasing with the shrinking of the area covered by ice [7,8] and there are an increasing number of cases of rising flood risks in coastal areas due to sea level rise [9]; on the other hand, we are experiencing increasing levels of drought-endangered areas and drought [10]. These changes in environmental conditions are also a catalyst of shifts in the geographical ranges of many trees and shrubs [11,12]. Long-lived organisms are most at risk of changes that (i) could modulate their metabolism and cause species to adapt to new conditions or (ii) could have an impact severe enough to cause the death of the organism. The changes in the climate are believed to greatly affect the state of worldwide arboreal flora. We must take into account the fact that tree species differ in their ability to adjust to new conditions and in the rate at which they are able to do so. Overall, changing climate introduces considerable shifts in selective pressure in tree populations, and these shifts may result in different distributions of individual species in the future [13,14].

We review and discuss various aspects of the modulation of tree physiology at the seed stage, at the germination stage and during further development stage, such as phenology, epigenetics and phenotypic plasticity, from the perspective of adaptation to rapidly changing climate. This work is an attempt to understand the pathway of changes that affect ecological connections between different species of trees and their rapidly changing environment. With this perspective, we attempt to predict how the current situation may evolve. This understanding may bring us closer to more adequately planning ways in which we can conquer the danger to biodiversity that arises from a changing climate.

2. Forest Threats Related to Global Warming

Due to increasing temperatures, the scale of the problem of global drought-induced stress and mortality of forest species is currently increasing at a rate unprecedented in recorded history [15,16]. During a 4-year extreme drought lasting from 2012 to 2015 in California, tree mortality increased from tens to hundreds of dead trees e.g., *Abies magnifica* (A. Murr), *Abies concolor* (Gord. & Glend.), *Lithocarpus densiflorus* (Hook. & Arn.) and *Pseudotsuga menziesii* (Mirb.) [17] per km^2. It is therefore important to determine whether the mortality of trees in drought-affected areas will be concentrated in specific competitive and climatic conditions because this determination will enable forest managers to effectively identify areas that are more vulnerable to drought. Large-scale, consistent monitoring of forest ecosystems plays a key role in preparing for the impact of extreme events on the stability of forests [18].

To predict the traits of plants that are likely to die or to survive during drought under the influence of a future warmer climate, Darcy's law, which describes the relationship between the instantaneous discharge rate through a porous medium, the viscosity of the fluid and the decrease in pressure over a given distance, can be used. Through application of this core principle of vascular plant physiology, it has been established that it is likely that tall plants with isohydric stomatal regulation, low hydraulic conductance and high leaf area as trees' species from Fagaceae and Sapindaceae families will die because of drought stress in the future [19]. The rise in temperature is a cause of increasing vapor pressure deficits, and the resulting water stress will result in changes in the plants' structure. Old-growth forest mortality is likely to increase in a warming climate, while low-statured plants will probably survive, and forest ecosystems will be gradually replaced by communities of shorter grasses or shrubs [20].

2.1. Direct Impact of Global Warming on Forests-Abiotic Factors

2.1.1. Responses at the Seed Level

The role of different stages of thermal history, including previous generations and stages of flowering and seed development, still requires careful investigation [21–25]. The results of some studies suggest that there is an interaction between provenance and climate that has a significant impact on successful development and survival and that phylogenetic signals do not play an important role in this adaptation [26]. Thermal signals shape the successive steps of seed production, dormancy and germination; however, due to phenotypic plasticity, those processes also have thermal memory that contains their thermal experience from the past [25,27]. Microclimate may be the factor that most affects seed germination and early seedling establishment [13,28–31].

An important factor in the distribution of plant species is their germination behavior, one of the earliest phenotypes expressed by plants. This process frequently undergoes natural selection before other traits become observable because it is expressed early [30]. At this phase of life, plants are most susceptible to environmental stress. Researchers [32,33] have found that it is necessary to include seed size, lifespan and germination base temperature when analyzing the thermal limits of species distributions.

Seeds of phylogenetically related *Caragana* species that have germinated under different temperature and moisture conditions have shown differences in germination strategies that are tightly linked to the ecological niche of the species. This outcome indicates that the germination process is driven not by phylogenetic signals but by environmental causes [30]. Moreover, sycamore maple (*Acer pseudoplatanus* L.) seed lots collected along a north-south gradient spanning 21° of latitude in Europe differed in maturity, and the level of maturity was influenced by the heat sum during seed development. Seeds from colder locations, for example, Scotland, germinated over a narrower temperature range and were more desiccation-sensitive than seeds from Italy and France, where the heat sum could be as much as two times higher [31].

Seeds of pendunculate oak (*Quercus robur* L.) from oceanic climates in West Poland required more humid and warmer conditions for epicotyl emergence, while in seeds from continental climates in East Poland, epicotyl emergence occurred both in cold-dry and warm-wet conditions. This observation shows that populations are adapted to their local environments [28]. Therefore, seed germination strategy is not a stable evolutionary trait within a family or genus [30]. Fremont cottonwood (*Populus fremonti* Wats.) trees from populations occurring in warmer, southern climates exhibited up to four times greater plasticity than populations adapted to cooler northern conditions. The value of phenotypic plasticity is correlated with the local climate. Trees experiencing hotter climates generally exhibited smaller shifts in an adaptive direction compared to trees transferred to cooler climates, which showed non-adaptive changes [29].

2.1.2. Biochemical and Physiological Plant Responses

Under conditions of poor water availability caused by dryness, plants experience water stress. In response, multiple signaling pathways and response mechanisms are activated to counteract water loss and allow the plant to adapt to the resulting dangers.

When the level of water in a plant decreases significantly, its cells begin to lose turgor, resulting in plasmolysis. On the macroscopic scale, this phenomenon manifests as shoots becoming flaccid and leaves wrinkling and folding. When only 20% of the original amount of water remains in the plant's cells (a state known as desiccation), the cell membrane begins to shatter and lose its selective properties. Phospholipids assemble in hexagonal structures instead of in a bilayer. Proteins begin to lose their native structures due to changes in the distribution of electrical charges caused by increased salt concentration [34–37].

Among the physiological changes that occur in plants during water stress, osmotic adaptation is worth noting. It depends on the production and accumulation of osmolytes such as proline, mannitol, glucose and fructose, which help maintain the water potential of the plant's tissues at a level that effectively counteracts the water loss. Accumulation of inorganic ions in vacuoles also helps overcome soil salinity, preventing water intake [34]. There is effective vacuolar sequesteration of Na+ in *Pongamia pinnata* L. roots. These roots are thus a kind of filter, preventing the accumulation of excess Na+ in the leaf [38].

Another type of response is increased production of particular phytohormones that affect metabolism in a way that counteracts water stress. For instance, abscisic acid (ABA) stimulates a number of physiological changes, including increased emergence of cilia on the epidermis, lower viability of generative organs, flower drop, stimulation of root growth and inhibition of shoot growth [39–42]. When these changes occur, water usage by above-ground organs is limited, and roots that are extended quickly have a greater likelihood of reaching more humid soil levels. Observations of this phenomenon are pointing to plasticity in root stress response [43]. On the other hand, research at Oak Ridge, USA has not confirmed that the tree root system clearly reacts to drought [44]. In trees, an additional strategy is leaf drop stimulated by ethylene [45,46]. The variant of this strategy called cladoptosis, which depends on dropping whole shoots instead of merely leaves, occurs in the species *Quercus*. This variant reduces the amount of metabolically active, water-consuming tissues and cuts off the main pathway of water loss through transpiration [47].

One of the most important defense strategies used by plants is limitation of transpiration. First, ion pumps are activated in the cells surrounding stomata. The osmotic effect results in a change in the cells' turgor due to water accumulation and in eventual closure of the stomata [48–50]. This change makes transpiration virtually impossible because the above-ground organs of land plants are covered by a protective wax cuticle. Another step in limiting water loss is increased production of the aforementioned wax [51]. It has been shown that leaf area and leaf production decline with increasing water stress. Leaf area and transpiration rates were lower in water stressed than in non-water stressed plants [52–54]. However, it turns out that leaf area alone could not determine the rate of transpiration, because in some cases small leaves transpire more than large leaves, due to differences in the size and density of stomata on the leaf surface [55]. In the case of baobab (Adansonia digitata L.), it was confirmed that the higher the temperatures at which trees live, the denser is the distribution of stomata on the leaf surface [56].

The main negative consequence of stomata closure is lockdown of gas exchange between the plant and its environment. Carbon dioxide, which is necessary for photosynthesis, cannot make its way into the leaf. This outcome leads to a significant decrease in the concentration of this compound in leaf cells within a short time. CO_2 deficiency leads to effective competition by O_2 for the active center of the Rubisco enzyme [57–59]. Rubisco is known for its tendency to recognize O_2 instead of CO_2 as a substrate for connection with ribulose 1,5-bisphosphate (RuBP), which begins the photorespiration pathway. The plant does not benefit energetically from this process and loses a significant amount of carbon, as CO_2 is one of the byproducts. This outcome lowers the overall efficiency of photosynthesis. This tendency manifests mostly when the CO_2 to O_2 ratio in cells is low. High temperature of the reaction environment is also a stimulating factor. Metabolic pathways aimed at eliminating the harmful glyoxylate formed in this process require hydrogen peroxide, resulting in greater production of this molecule in peroxisomes [60,61]. Forest trees have C3 metabolism, thus, competition between substrates at the Rubisco level, which is dependent on temperature, with the ratio of oxygenation and carboxylation increasing with increasing temperature [59]. In the case of *Quercus pubescens* Willd., photosynthesis may be temporarily affected by a heat-dependent reduction of Rubisco activation state. Therefore, *Q. pubescens* is an example of species which may be negatively affected by the heat waves, which occur currently and could occur in the near future [62].

In the context of the light phase of photosynthesis, carbon dioxide deficiency limits the regeneration of the oxidized form of nicotinamide adenine dinucleotide phosphate (NADP). Accumulation of the reduced form of this nucleotide results in blockage of the electron transport pathway in the thylakoid membrane. The persistent reduced state of the chain complexes results in disruption of the balance of chloroplasts' redox state. A side effect is oxygen reduction on photosystem I in the Mehler reaction [61,63,64]. This reaction is often incomplete, resulting in the formation of a superoxide anion radical and hydrogen peroxide [59,64–66]. To balance the redox potential of the cell, the NADPH reduction potential is transported through a system of molecular "shuttles" to the cytosol and then to mitochondria, where it is used to feed the respiratory chain [67–69]. Excessive reduction of protein chain complexes, in particular complex III, in a manner analogous to that occurring in chloroplasts results in the formation of superoxide anion radicals [70–72]. In this context, the plant's defensive action is to increase the expression of alternative oxidase (AOX) [73–76]. This enzyme acts as a "safety valve" to eliminate the accumulating reductive force and thus limit the production of reactive oxygen species (ROS) [77].

Heat leads to denaturation of the protein components of cell membranes because it accelerates the movement of lipids. In effect, it makes cell structures more sensitive, which leads to the disruption of physiological and cellular processes such as ion transport and photosynthesis [78–80]. Concurrently, heat increases the formation of ROS, as which superoxide radical ($O_2^{-\bullet}$), hydrogen peroxide (H_2O_2) and hydroxyl radical ($\cdot OH$), are classified (Figure 1). Notably, studies on tree species of Dipterocarpaceae family have shown that species which occur at uplands and riparian fringes with a higher frequency of disturbances, are suggested to have higher photosynthetic tolerance to elevated temperatures [81].

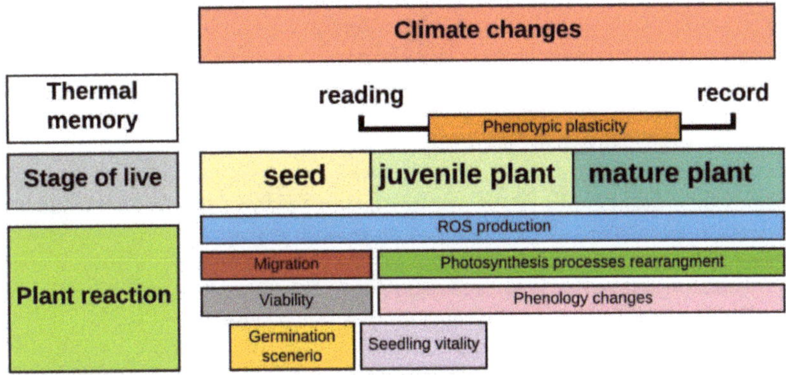

Figure 1. Rapid changes in climate conditions influence each of stage of a plant's life, from seed production by juvenile and mature plants. Thermal memory recorded due to phenotypic plasticity in mature organisms is passed on to the seed, in which the thermal memory is read. As a consequence of climate changes, reactive oxygen species ROS production may change at each of stage of life; important differences in the proper functioning of photosynthesis also occur. Acceleration or delay of developmental programs is observed as a change in the phenology of the species. Changes in the climate may interfere with the viability of seeds and may cause modifications in germination scenarios or seedling vitality. Seeds are plant organs that may, as a consequence of these events, migrate to establish new territory and change the range of populations and even species.

These molecules are characterized by extremely high reactivity. They easily react with cell-building macromolecules, mainly in the mitochondria, chloroplasts and peroxisomes [59,82,83] disrupting their structure and function [84]. The loss of seed viability during storage at temperatures above 0 °C is closely linked to ROS production, and the antioxidant system is not always sufficient to protect them [85]. ROS induce oxidative damage to nucleic acids, lipids and proteins, thereby affecting organelles and their metabolic integrity [86–91]. These reactions include lipid peroxidation (LPO) (a process of oxidation of polyunsaturated fatty acids by $O_2^{-\bullet}$) [92], oxidation of protein disulfide bonds and peroxidation of lipids and proteins by H_2O_2 [36,93,94]. Excessive accumulation of reactive oxygen species causes irreversible damage of the cell and leads to cell death [90,95]. The presence of ROS stimulates the defense response in the form of increased activity of enzymes that catalyze the breakdown of harmful substances into less reactive forms. These include peroxidases, which catalyze the conversion of hydrogen peroxide to molecular oxygen and water, and superoxide dismutase (SOD), which catalyzes the decomposition of anion radicals to less harmful H_2O_2 [96,97]. A number of small molecular substances such as reduced glutathione, tocopherol and ascorbic acid also participate in these reactions. Tocopherol is anchored in the cell membrane, where it protects lipids from peroxidation. With the participation of appropriate peroxidases, ascorbic acid and glutathione are oxidized during the H_2O_2 decomposition reaction, thereby decreasing the overall quantity of these molecules in the cell [98,99].

2.2. Indirect Impact of Global Warming on Forests-Biotic Threats

Climate change is already influencing the geographic distribution of pathogens and insects that prey on tree species. The changes facilitate the expansion not only of exotic pests but also of native pests. In addition, climate change can affect tree resistance to pests, and this is an increasingly frequent phenomenon [14,100]. Temperature is a main factor affecting many of the insect life-history events on which population success depends [101]. Climate change contributes to shorter generation times of forest insect species. For example, an increase in temperature can facilitate survival of winter conditions by the mountain pine beetle (*Dendroctonus ponderosae* Hopkins) [102], which is the most serious insect pest of western North American pine forests, and its outbreak a few years ago caused the destruction of over 15 million hectares of pine forests. This event had a significant impact on forest health and the economy of the woodworking industry [103]. Moreover, drought is one of the most important climatic factors affecting host trees' susceptibility to forest insects because reduced water availability threatens the resilience and viability of trees [100]. The simultaneous occurrence of invasive insect outbreaks in forests and the weakening of trees' resistance to their attacks results in increased tree mortality. Drought is thought to be the main cause of infestation by several defoliators and bark beetle species, as well as pathogens [9,104]. For example, susceptibility to occupation of Norway spruce stands in Poland by pests is relatively high because this tree species possess a shallow root system; therefore, it experiences drought stress more rapidly than deep-rooted species, and at the same time, the production of resin, which is used to repair injuries, including insect bites, is reduced [105,106]. As a consequence, the number of bark beetles (*Ips typographus* L.) is changing rapidly, resulting in a serious problem of massive dying-off of Norway spruce stands in northeastern Poland and, at the international scale, infecting areas such as the Białowieża Forest [107]. Since 2012, spruce stands covering a total area of approximately 1.4 million m^3 have died (www.bialystok.lasy.gov.pl). The development and reproduction of the bark beetle population are stimulated by factors such as high temperature and low precipitation. An increase in the number of bark beetles, a species with high reproductive potential that can produce two main generations and one or two sister generations in a year, usually leads to catastrophic effects and significant economic losses throughout the area occupied by Norway spruce [108]. Climate change may also be a reason for an increase in the harmful activity of species that previously did not have significant economic significance in a given area. The main factor limiting the occurrence of *Nodiprion sertifer* (Geoffr.) in northern Finland is the occurrence of days with a temperature lower than −36 °C in winter; below this temperature, high egg mortality is recorded. In the same research, the scenario that would be expected to occur following an increase of 3.6 °C in the average winter temperature by 2050 was considered. Climate warming may lead to an increase in the frequency of gradation of tropical blue bees in areas where sporadic or non-occasional occurrence is presently found [109]. The impacts of climate change on forest insect and pathogen populations is an important research focus because management of the impact of invasive exotic and native pests becomes increasingly difficult [110]. Changes in the ranges of insects resulting from the impact of changed climate parameters may also result in their adaptation to new food plants. When species closely related to the original food plant are present in the new area of phytophage occurrence, it is possible that pests will broaden the spectrum of the host plants they feed on [111]. An increase in the average temperature over several decades has resulted in the expansion of the pine processionary (*Thaumetopoea pityocampa* Schiff.); to date, this expansion has been observed in the Sierra Nevada mountains of southeastern Spain, above the free parts of the mountains and extending into the areas of the mountains far beyond. The change in distribution was also accompanied by adaptation to a new host plant, a relict subspecies of the common pine, *Pinus sylvestris* var. *nevadensis* [112].

Fungi forest pathogens represent another important group of invasive organisms [113], and the major pathway for introduction of these pathogens is human-mediated transport [114]. An example of a common forest disease in Europe is oak powdery mildew [115], which is caused by a few species of *Erysiphe*. Epidemics of this pathogen have been observed in several years. According to the latest research, the increase in temperature related to greenhouse gas emissions has resulted in a marked increase in winter precipitation in northern Eurasia and eastern North America since 1920 [116]. The results of one study [117] showed that illness peaks occurred only after mild winters, although it was believed that climatic factors during the growing season have a greater impact on disease dynamics. The authors of this study also showed that there are evident interactions between powdery mildew infection and oak growth dynamics. It is possible that temperature conditions that occur in winter favor the perennation of powdery mildew fungi [118]. Moreover, cold winters limit overwintering of powdery mildew mycelia in leaves and buds [119] and cause limited survival of chasmothecia [120,121].

A loss of biodiversity can be indirectly related to climate, as it may be caused by shifts in the habitats of invasive species. In Western Europe, there are reports of potential invasion by several tree species from North America. These species include *Quercus rubra* L. and *Robinia pseudoacacia* L. The invasive potential of these species, favored by changing climate, is expected to jeopardize the biodiversity of the invaded regions. As of the writing of this document, the area of Europe harboring hospitable conditions for these species is thought to be halfway occupied [122].

3. Novel Environmental Conditions as Changes Generator

Tree species living in rapidly changing climate conditions are exposed to danger and must adapt to new climate conditions to survive. Trees respond to changes in the environment in numerous ways. Physiological modulation at the seed stage, germination strategy and further development are influenced by many different factors [13,14].

Research by [123] on Norway spruce (*Picea abies* L.) shows that the adaptive abilities of particular tree species can be surprisingly high. The authors discovered a form of adaptation that is passed to the offspring and is based on epigenetic memory. The length of day and temperature conditions under which parent plants were maintained influenced the growth of their progeny during the following year. This influence indicates that in some cases, trees are able to adapt surprisingly quickly to a changing environment (Table 1).

A warming climate results in earlier leaf unfolding in spring for many tree species, increasing the risk of exposure of juvenile leaves to frost damage. [126] examined 13 species of trees and shrubs in Switzerland and compared the measurement results obtained over a 60-year period. There was a noticeable trend of leaf unfolding occurring earlier in later years that correlated with the increase in global temperatures.

Lankau et al. [127] explored how climate change affects trees' production of mycorrhiza, with distinction between ectomycorrhiza and arbuscular mycorrhiza. Both types of symbiosis are observed to mitigate the sensitivity of trees to changing climate. Further research is needed to verify the deeper connections between changing climatic conditions and the relationships between trees and their symbiotes.

Another study [128] described the relationship of pinyon pine (*Pinus edulis* Engelm.) and its ectomycorrhizal fungal communities (EMFs). Pinyon pines with particular genotypes are observed to associate only with particular EMFs. Additionally, mycorrhizal fungi of different species provide different types of enhanced resistance for their tree counterparts. Both drought-resistant and drought-resilient species of plants showed signs of improved drought survival when associated with fungi of the genus *Geospora*. Affinity for specific fungal symbionts is part of a tree's phenotype and may serve as a form of adaptation.

Populations of trees of the same species may differ in phenotype. *Fagus sylvatica* (L.) in populations located in the core of the species' aerial range appeared to be the most susceptible to intense drought and suppressed growth. On the other hand, the highest durability was observed in populations located in the edges of their habitat close to the equator. This observation leads to the conclusion that future changes in temperature and overall climate might effectively compromise the range of this species, which is currently dominant in Europe [129].

Table 1. Diversifying the forms of adaptation of selected tree species to changing climatic conditions.

Stage of Development	Species	Type of Adaptation	References
Seed stage	*Quercus petraea* (Matt.) Liebl. *Quercus robur* L.	Increasing temperature improve seed resistance to environmental stress and increase germination efficiency	[124]
Germination stage	*Caragana* ssp.	Germination strategies tightly linked to the ecological niche of the species	[30]
	Quercus robur L.	Offspring of the population occurring in the continental climate has a wide range of tolerance to abiotic environmental conditions	[28]
Further developmental stages	*Picea abies* L.	Epigenetic "memory": impact of simulated length of day and temperature, in which parent plants were kept, on the growth of their progenies	[123]
	Populus fremonti Wats.	Offspring of the populations occurring in warmer conditions shows four times greater plasticity, unlike the populations that adaptedto cooler conditions	[29]
	Picea mariana Mill.	The wide range of abiotic conditions in which the species occurs causes the formation of populations that vary in terms of specific adaptations to local conditions	[125]
	Pseudotsuga menziesii Mirb.	Growth initiation on Douglas-fir will track progressive changes in favorable climatic conditions at high elevations and latitudes	[12]

Trees are used as a popular countermeasure to reduce excessive heat in urban areas. This effect includes providing shade for trolley surfaces and cooling through transpiration. According to [130], specimens of *Quercus ilex* (L.) examined in four different urban sites were exposed to excessive water stress, which resulted in xylem cavitation, increased embolism and stomatal aperture. Under a continuous trend of summer heat waves, urban trees are at risk of dieback. On the other hand, there is some evidence that trees grow faster in urban areas. The rate of growth of urban trees appears to have increased since the 1960s, especially in cities located in the boreal climate zone [131]. It seems to be related to the urban heat island effect, which probably stimulates photosynthetic activity and extends the growing season [132]. However, in temperate climate, urban tree growth is not as significant. Trees probably suffer from substantial water stress, due to, e.g., impervious surfaces and compacted soils in urban areas, which may reduce root growth and in turn hamper a tree's water uptake [131].

Climate Is an Important Selective Factor in Populations

The plasticity of traits such as those discussed above is under the influence of evolutionary change and affects the long-term persistence of populations and their roles in ecosystems [133]. Many species are not able to migrate rapidly enough to keep up with the rapidly changing climate; therefore, adaptation must play a key role in their response [134]. In fragmented habitats, rapid climate change probably surpasses the capacity for adaptation in many plant populations, and their genetic composition may be dramatically altered as a result [135]. It is possible that the consequences will include unpredictable changes in the number of species within communities and a reduction in their ability to survive further environmental perturbations such as pest outbreaks, disease outbreaks and extreme climatic events [136].

In the near future, an increase in the frequency of extreme climatic events is predicted; thus, individuals would need to possess practically perfect plasticity to allow them to tolerate all fluctuations in climate conditions with no fitness losses [137]. The distributional changes and climate-related forest dieback that are currently taking place show that such universal plastic tolerance of the changing climate is not standard [138–141]. A plastic phenotype will enable plants to respond to climate changes during the lifetime of the individual (Figure 1); however, the capacity for a plastic response to an event is decreasing as more extreme events of greater extremes experienced and are lasting longer [136]. Therefore, the capacity of plants to respond to environmental changes by phenotypic plasticity has limits [134,142].

Tree species that are locally adapted at their limits of range could be beneficial for range expansion because, given that environmental conditions in new areas are similar to the current ones, colonizing genotypes are the fittest. If climate changes occur faster than trees' acquisition of the capacity to migrate, local adaptation can make their range expansion slower [13]. European beech (*Fagus sylvatica* L.) is an example of an expansive tree species, and ongoing changes in climate conditions could expand its potential range. Beech stand vitality is high, and its regeneration is well adapted to different site conditions at the limit of the species' distribution; thus, European beech will probably reach its northeastern limit in the future [143].

4. The Impact of Phylogeny on Plant Phenology

Despite the evidence that plant responses depend on abiotic conditions, it still has not been fully confirmed that seed germination, and consequently the distribution of species that evolved from the same ancestor, was formed under the influence of environmental conditions [13,28–30,144,145]. Some studies have found that tree phenology is determined by phylogenesis. When tree species from seven different biomes in the world were compared, species in the Arctic region were found to have a thermal upper limit lower than that of species in the rainforests. The maximum temperature difference for photosynthesis and respiration between species from Alaska and Peru was 9 degrees Celsius [146]. Among other findings, it was confirmed that species in North American forests have germination strategies similar to those of related species present in East Asia [147]. There is evidence that phylogenetic relationships are the factors that shape seed mass evolution. An increase in seed production, observed over the last decade in oaks in the temperate zone, increases seed dispersal [148]. Moreover, increasing temperature favors acorn mass increases (in case of *Quercus petraea* (Matt.) Liebl. and *Quercus robur* L., by about 0.15g per degree [124], which can improve seed resistance to environmental stress and increase germination efficiency [149].

As a result of phylogenetic limitations, related species do not spread far from their optimal ecological niches, and they also exhibit similar seed germination strategies [32]. Numerous studies investigating the importance of phylogeny in determining responses to climate change have been published [150]; however, most of these studies address herbals [151,152] and plant life-history events such as flowering [153,154] rather than trees or seed germination. A comparative analysis of phenological traits showed that the timing of life-history events covaries with phylogeny such that more closely related species tend to flower and leaf at similar times. Flowering time is a particularly conserved feature within tropical and temperate phylads [155]. It is possible that the phenology of flowering and seed development exhibits thermal memory. Flowering shows response to temperatures experienced comparably early in the maturation stage [156]. It has also been shown that seed weight depends on the duration of the flowering process [157]. Within natural populations, relatively rapid evolutionary changes can occur; however, the rate of temperature increase, which determines development, can be too rapid for populations to evolve in the absence of gene flow from populations of *Betula* with earlier dates of budburst [158]. Observations of the effects of temperature during ripening on the future features of the germination process suggest that plants' phenological thermal memory allows seeds to adapt their germination phenology to climate change. This phenomenon can help predict climate change [25].

Most published results indicate a positive effect of temperature increase on the seed germination process [159], but an experiment on 17 tree species in elevated temperature and CO_2 concentration [160] showed that long-term exposure of the seeds to high temperature is probably the main cause of the decrease in seed germination. Moreover, high temperatures are the cause of thermo-dormant changes, which are variable depending on the genotype; thus, temperature-dependent hormones and enzymes may be inactivated [161]. After all, in many cases, unlike in the above-mentioned study, high temperature breaks seed dormancy and stimulates initiation of germination [162]. It cannot be excluded that global warming reduces the amount of chilling trees experience, and then more forcing may be required to start growth [163], wherein it causes changes in the timing of growth initiation to lag behind climate change [164]. Many tree species require chilling to start budburst [161] but some studies suggest that the observed trend toward earlier budburst will be reversed if winter temperatures increase substantially [165]. To determine if changes in phenology are likely to reflect climate change, ref [12] modeled related shifts in tree phenology. For this purpose, they analyzed the time of initiation of growth and diameter growth in coast Douglas fir (*Pseudotsuga menziesii* Mirb.), which is a species of ecological and economic importance in western North America. The created models showed that growth initiation in Douglas-fir will track progressive changes in favorable climatic conditions at high elevations and latitudes, but on the other hand, it will lag behind these shifts at lower latitudes and elevations [166]. Diameter-growth initiation does not appear to require some chilling, but it may cause shifts in the diameter-growth initiation behind climate change over a larger geographical area, compared to the height-growth initiation. It is therefore really important to analyze many growth-initiating events when studying the relationship between climate and tree growth phenology [167].

5. Phenotypic Plasticity Allows Species to Survive under Novel Conditions

Phenotypic plasticity is defined as the ability of an organism to change in response to stimuli or resources from the environment. The response may or may not be adaptive; if adaptive, it is called adaptive plasticity. It may involve a change in morphology or physiological state or both at any level of organization, wherein the phenotype consists of all of the characteristics of an organism other than its genes [168].

Natural populations with high levels of genetic variation are characterized by increased potential to adapt to abiotic and biotic environmental changes [168]. The populations adapt through natural selection or migrate to follow conditions to which they were initially adapted and, notably, these options are not mutually exclusive [169,170]. Plant species can adjust to changing environmental conditions through phenotypic plasticity. It is likely that by using discoveries in the field of molecular biology, which is constantly developing, we can understand the mechanisms that underlie phenotypic plasticity. There is evidence that such plasticity is hereditary and genetically controlled and that it contributes to evolution [171,172].

The role of epigenetics in the process of phenotypic plasticity remains unclear despite recent discoveries suggesting that it is of great importance [173,174]. Studies investigating the role of epigenetics in tree phenotypic plasticity concluded that epigenetic mechanisms in forest trees play a crucial role in plants' ability to withstand and adapt to stressful conditions. These mechanisms probably mediate phenotypic modifications that result in beneficial effects in response to the environment. Changes in DNA methylation may play a key role in phenotypic plasticity [175]. In 2010, a possible link between epigenetics and plasticity in response to drought stress in poplar was investigated [176], and it was possible to establish correlations between morphological and epigenetic variables. Correlations between DNA methylation levels and biomass productivity have been demonstrated in a collection of poplar genotypes grown outdoors under various moisture conditions [177]. Moreover, differences in transcript abundance levels under the influence of stress have been reported. The observed differences varied depending on the geographic provenance, leading to the assumption that DNA methylation may lead to phenotypic variation induced by stress factors through modulation of gene expression [178]. Variations in soil water availability preferentially drive changes in the DNA methylation levels of genes involved in phytohormone metabolism and signaling; these changes potentially promote phenotypic plasticity [179]. The high level of phenotypic plasticity found in stone pine trees is probably linked to epigenetic variation between individuals or populations [180]. This species is characterized by very low levels of genetic variation and high levels of plasticity for many traits. Using epigenetic markers, it was possible to discriminate between two populations that were indistinguishable using classical molecular markers [181].

Long life span is a trait of many species of forest trees, and processes such as growth, bud bursting time and resistance to frost may be subject to climatic conditions [182,183]. Norway spruce (*Picea abies* L.) is the tree species in which epigenetic memory has been most researched. It was shown that temperature-dependent epigenetic memory affecting the timing of bud burst and bud set in trees is generated by temperature changes during somatic embryogenesis [123]. Stem cuttings from poplar derived from populations located in areas with different amounts of phosphorus developed differently in the same habitat, despite their clonal origin. These differences in seedling growth strategy were associated with different levels of methylation of miRNAs as a function of primary environmental conditions, and it was a cause of habitat-dependent root modification under conditions of phosphate starvation [177].

6. Future Perspective for Research and Practice Solution in Terms of Adaptation of Trees to a Rapidly Changing Climate

6.1. Preventing Harmful Climatic Events by Forecasting the Responses of Tree Species

Range reduction could have serious consequences for nature protection and forest management [11]. Assisted migration cannot be used without determining the future climatic conditions. By employing databases on the current distributions of tree species and the latest climate scenarios, it is possible to predict the changes in the ranges of species that are likely to occur in the next decades [184].

Many models that address the current and predicted distribution of tree species have been developed. These models are based on different data sets and take into account different variants of future events [11]. Moreover, further properties were found that may be affected by species boundaries, and these may prove valuable in forecasting tree species' mortality. For example, a study conducted in the United States showed that hydraulic traits can influence species distributions and that knowledge of these traits may be valuable in forecasting drought-induced tree mortality [14]. A model developed as part of this study captured the high prevalence of aspen mortality in the western and southern limits of aspen distribution in this region. Modeling of climate space suitability for the Dipterocarp trees Sal (*Shorea robusta* Gaertn.) and Garjan (*Dipterocarpus turbinatus* Gaertn. F), which play an important role in the economies of many Asian countries, showed that annual precipitation is the main variable that explains the current and future distribution of these species [185]. The results of this study can be used to identify sensitive species habitats and in designing appropriate conservation activities.

Reference [11] identified areas in which potential climatic conditions will be optimal for the tree species most important in forest management in Europe and areas in which these species may be threatened in the future. This study showed that species that will increase their suitable habitat areas as a result of climate change are likely to be mostly late-successional species such as European silver fir (*Abies alba* Mill.) and European beech (*Fagus sylvatica* L.), while the distribution ranges of most pioneer species such as Scots pine (*Pinus sylvestris* L.) and silver birch (*Betula pendula* Roth) would decrease (Figure 2) [186].

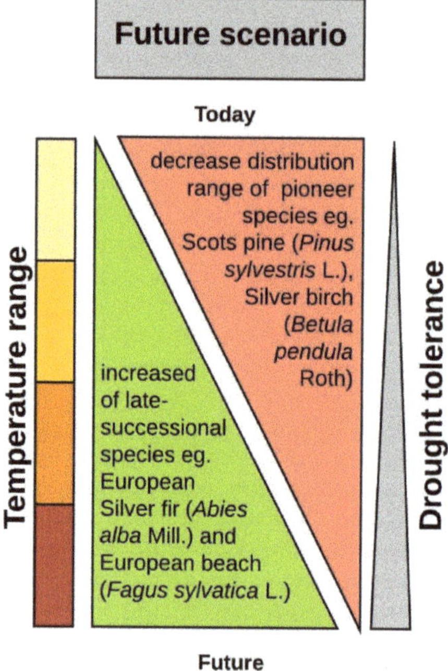

Figure 2. Diagram illustrating future scenario of tree species redistribution according to differences in the temperature range and drought tolerance of species grouped as pioneer and late-successional.

These findings show that the distribution of the studied species is mainly limited by water deficit during the growing season caused by more intense transpiration at higher temperatures [11]. In addition, Reference [27] indicated that models of future dispersal should consider how future climate conditions will impact seed number and mass, because these traits determine the dispersibility of individuals and populations and the health of their seedlings.

6.2. When Adaptation Is not Fast Enough—Assisted Migration

One solution that has been developed to mitigate the consequences of climate change is assisted migration [6]. Assisted migration relies on the purposeful movement of species to imitate or facilitate natural range expansion [187]. Assisted migration might be useful for locally adapted and long-lived populations or species, especially when their migration capacity is limited or when their habitats are threatened by fragmentation [188,189]. Using genotypes of Fremont cottonwood (*Populus fremontii* Wats.) from 16 populations from throughout the species' thermal range, it was shown that the beneficial plasticity of northern populations moved to hotter regions made it possible for these populations to experience relatively large amounts of warming before they reached a mortality threshold. In contrast, southern populations moved up in elevation and latitude showed non-adaptive, unfavorable plasticity responses [29]. Consequently, active management of the areas by methods such as assisted migration is truly needed. Climate change pushes the frost-line north; thus, the best solution, especially in areas in which winters are relatively warm, might be planting genotypes from lower elevations or latitudes adapted to long growing seasons. Long-lived trees can therefore benefit from plasticity to adapt to a rapidly changing environment [29,190].

It will be necessary to match seed sources to the climates of the next decades to ensure the survival of forest tree species. Short-distance transfers could be used to buffer uncertainty about the scale of climate change in an area by improving the gene flow among populations by planting more diverse seed sources among and within forest stands [191,192]. Shifting climates may even render current species or populations non-adapted.

Understanding the process that can lead to production of refugia from contemporary climate change would be advantageous in preserving genetic diversity [193]. Because assisted migration certainly is not a solution consistent with our previous understanding of natural resource management or with previous views in conservation biology, it must be realized in such a way as to include valuation of species and population vulnerability to climate change, selections of options and management targets and long-term monitoring [186]. Such proceedings can become effective, inexpensive and practical options for addressing climate change. Currently, local assisted migration initiatives are underway in Canada and British Columbia [193].

7. Conclusions

Changes in climate conditions lead to global drought-induced stress and increase plant mortality. The impact of climate change is becoming much more noticeable from year to year. Nowadays, it is very important to monitor forest ecosystems, and each change should be described as a function of time and in relation to other factors, e.g., temperature, rainfall, etc. Collecting data on the changes in the phenology of species, changes in the related pathogens, insects and animals (number of species and time of settlement) is of real value in predicting the rates of change and their consequences. These changes may be manifested on a small scale as fragmentation and redistribution of populations, changes in range, and the development of adaptation; when the changes are too rapid or their impact is too great, they may be reflected in the death of organisms/populations, leaving ecological niches empty (Figure 3).

For proper interpretation of these changes, we should use a systems biology approach. Creation of a database that would compile observations from different levels of organism systematic versus changes in their ecological behavior and physiology may bring us closer to a comprehensive interpretation of the changes that occur due to rapidly changing climatic conditions.

Rapid climate change causes modifications in the physiology of trees at various levels ranging from molecular and biochemical changes to plant structural changes. Drought alone or drought followed by flood results changes in osmotic adaptation and in the accumulation of proline, mannitol, glucose and fructose, substances that play a key role in maintaining the water potential of plant tissues. Furthermore, drought leads to changes in the functioning of the photosynthetic apparatus that are reflected as stomata closure, carbon dioxide deficiency and limiting of transpiration. On the other hand, heat stress causes the denaturation of proteins in cell membranes and increases ROS production (Figure 3).

In spite of research results, which are not conclusive, it is possible that phenotypic plasticity has an important role in plants' adaptation to environmental changes. This finding places the issue of proper seed reserve conservation in gene banks in a new light. Reserves with the highest genetic diversity and from populations developed in different locations (with differences in adaptation to microclimate conditions) have the best opportunity to survive under new climate conditions and during changes in ecosystems that we cannot predict today. Epigenetic studies conducted in the next few years may bring us closer to understanding the mechanisms through which trees adapt to stressful conditions, especially in the context of epigenetic memory in long-lived organisms. The question "how does epigenetic memory develop according to the thermal history of the plant?" is still open.

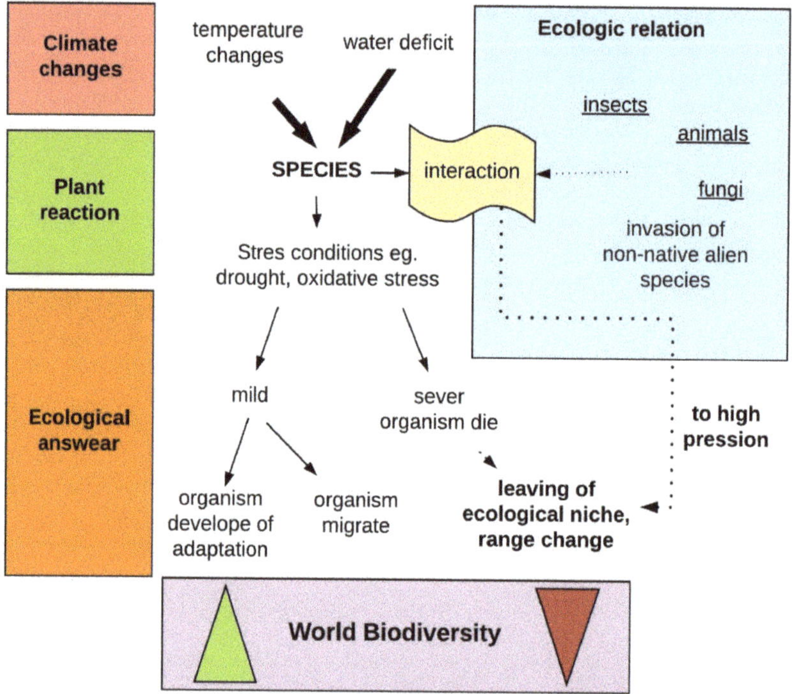

Figure 3. The relationship of climate change, plants' reactions and ecological behavior to world biodiversity. Rapid changes in climate (e.g., temperature and water status) influence species, and plant physiology changes in response to stress conditions. If stress conditions are mild, the organism may develop adaptation and protect the range of the population (increase in biodiversity), or migrate (no changes in biodiversity but changes in the distribution of species); severe stress may lead to death of the organisms and, in perspective, to an ecological decrease in world biodiversity. In ecosystems, plant species interact with other organisms such as fungi, animals, insects and invasive nonnative alien species. This interaction may aggravate stress, and when the pressure becomes too high, plant species may leave their niches, changing the range of species, and a decrease in world biodiversity would be observed.

Many scenarios and models based on different data sets and aspects have been constructed. We are still far from understanding how rapidly climate changes switch over plants from different biomes due to different strategies of plant population for microevolution, epigenetic memory and thermal history. Drought tolerance and adaptation during each stage of plant development are still considered to be main trait for developing future scenarios for forestry in the northern hemisphere. According to findings based on the drought tolerance of various species, mature late-successional species such as European silver fir (*Abies alba* Mill.) and European beech (*Fagus sylvatica* L.), will increase their range, in contrast to mostly pioneer species such as Scots pine (*Pinus sylvestris* L.) and silver birch (*Betula pendula* Roth), whose distribution may decrease (Figure 2).

Author Contributions: Conceptualization J.K.-O. and E.R.; Investigation, E.R.; Writing—Original Draft Preparation, J.K.-O. and J.K.; Writing—Review and Editing, E.R. and A.M.S.; Visualization, E.R. and A.M.S.; Supervision, E.R.; Funding Acquisition, E.R. All authors have read and agreed to the published version of the manuscript.

Funding: This research was supported by the National Centre of Sciences (grant number 2018/31/B/NZ9/01548). Additional funding was provided by the Institute of Dendrology, Polish Academy of Sciences.

Conflicts of Interest: The authors declare that they have no conflict of interests.

References

1. Ebele, N.E.; Emodi, N.V. Climate Change and Its Impact in Nigerian Economy. *J. Sci. Res. Rep.* **2016**, *10*, 1–13. [CrossRef] [PubMed]
2. Costinot, A.; Donaldson, D.; Smith, C. Evolving Comparative Advantage and the Impact of Climate Change in Agricultural Markets: Evidence from 1.7 Million Fields around the World. *J. Political Econ.* **2016**, *124*, 205–248. [CrossRef]
3. Stolpe, M.B.; Medhaug, I.; Knutti, R. Contribution of Atlantic and Pacific Multidecadal Variability to Twentieth-Century Temperature Changes. *J. Clim.* **2017**, *30*, 6279–6295. [CrossRef]
4. Anderson-Frey, A.K. *Statistical Examination of Tornado Report and Warning Near-Storm Environments*; The Pennsylvania State University: University Park, PA, USA, 2017; ISBN 9780355329568.
5. Vogel, M.M.; Zscheischler, J.; Wartenburger, R.; Dee, D.; Seneviratne, S.I. Concurrent 2018 Hot Extremes Across Northern Hemisphere Due to Human-Induced Climate Change. *Earth's Future* **2019**, *7*, 692–703. [CrossRef]
6. Wang, T.; O'Neill, G.A.; Aitken, S.N. Integrating environmental and genetic effects to predict responses of tree populations to climate. *Ecol. Appl.* **2010**, *20*, 153–163. [CrossRef]
7. Jakobson, L.; Vihma, T.P.; Jakobson, E. How does the shrinking sea ice influence changing wind speed over the Arctic Ocean? Presented at the AGU Fall Meeting, San Francisco, CA, USA, 12–16 December 2016.
8. Chambers, A. *The Arctic's Shrinking Sea Ice Gives Way to Increasingly Navigable Shipping Corridors: GIS Based Analysis by Satellite Imagery and Radar Observation*; Western Washington University: Washington, DC, USA, 2017.
9. Wdowinski, S.; Bray, R.; Kirtman, B.P.; Wu, Z. Increasing flooding hazard in coastal communities due to rising sea level: Case study of Miami Beach, Florida. *Ocean. Coast. Manag.* **2016**, *126*, 1–8. [CrossRef]
10. Young, D.J.; Stevens, J.T.; Earles, J.M.; Moore, J.; Ellis, A.; Jirka, A.L.; Latimer, A.M. Long-term climate and competition explain forest mortality patterns under extreme drought. *Ecol. Lett.* **2017**, *20*, 78–86. [CrossRef]
11. Dyderski, M.K.; Jagodziński, A.M. Drivers of invasive tree and shrub natural regeneration in temperate forests. *Boil. Invasions* **2018**, *20*, 2363–2379. [CrossRef]
12. Bigler, C.; Bugmann, H. Climate-induced shifts in leaf unfolding and frost risk of European trees and shrubs. *Sci. Rep.* **2018**, *8*, 9865. [CrossRef]
13. Solarik, K.A.; Messier, C.; Ouimet, R.; Bergeron, Y.; Gravel, D. Local adaptation of trees at the range margins impacts range shifts in the face of climate change. *Glob. Ecol. Biogeogr.* **2018**, *27*, 1507–1519. [CrossRef]
14. Anderegg, W.R.L.; Flint, A.; Huang, C.-Y.; Flint, L.; Berry, J.A.; Davis, F.W.; Sperry, J.S.; Field, C.B. Tree mortality predicted from drought-induced vascular damage. *Nat. Geosci.* **2015**, *8*, 367–371. [CrossRef]
15. Berg, N.; Hall, A. Increased Interannual Precipitation Extremes over California under Climate Change. *J. Clim.* **2015**, *28*, 6324–6334. [CrossRef]
16. Williams, A.P.; Seager, R.; Abatzoglou, J.T.; Cook, B.I.; Smerdon, J.E.; Cook, E.R.; Williams, P. Contribution of anthropogenic warming to California drought during 2012–2014. *Geophys. Res. Lett.* **2015**, *42*, 6819–6828. [CrossRef]
17. Eyre, F.H. *Forest Cover Types of the United States and Canada*; SAF: Washington, DC, USA, 1980; ISBN 978-0686306979.
18. De Andrés, E.G. Interactions between Climate and Nutrient Cycles on Forest Response to Global Change: The Role of Mixed Forests. *Forests* **2019**, *10*, 609. [CrossRef]
19. McDowell, N.G.; Allen, C.D. Darcy's law predicts widespread forest mortality under climate warming. *Nat. Clim. Chang.* **2015**, *5*, 669–672. [CrossRef]
20. Williams, A.; Allen, C.D.; Macalady, A.K.; Griffin, D.; Woodhouse, C.A.; Meko, D.M.; Swetnam, T.W.; Rauscher, S.A.; Seager, R.; Grissino-Mayer, H.D.; et al. Temperature as a potent driver of regional forest drought stress and tree mortality. *Nat. Clim. Change.* **2013**, *3*, 292–297. [CrossRef]
21. Fernández-Martínez, M.; Vicca, S.; Janssens, I.A.; Espelta, J.M.; Peñuelas, J. The North Atlantic Oscillation synchronises fruit production in western European forests. *Ecography* **2017**, *40*, 864–874. [CrossRef]
22. Koenig, W.D.; Knops, J.M.H.; Pesendorfer, M.B.; Zaya, D.N.; Ashley, M.V. Drivers of synchrony of acorn production in the valley oak (*Quercus lobata*) at two spatial scales. *Ecology* **2017**, *98*, 3056–3062. [CrossRef]
23. Vacchiano, G.; Ascoli, D.; Berzaghi, F.; Lucas-Borja, M.E.; Caignard, T.; Collalti, A.; Mairota, P.; Palaghianu, C.; Reyer, C.P.; Sanders, T.G.; et al. Reproducing reproduction: How to simulate mast seeding in forest models. *Ecol. Model.* **2018**, *376*, 40–53. [CrossRef]

24. Bogdziewicz, M.; Szymkowiak, J.; Fernández-Martínez, M.; Peñuelas, J.; Espelta, J.M. The effects of local climate on the correlation between weather and seed production differ in two species with contrasting masting habit. *Agric. For. Meteorol.* **2019**, *268*, 109–115. [CrossRef]

25. Fernández-Pascual, E.; Mattana, E.; Pritchard, H.W. Seeds of future past: Climate change and the thermal memory of plant reproductive traits. *Biol. Rev. Camb. Philos. Soc.* **2019**, *94*, 439–456. [CrossRef] [PubMed]

26. Desnoues, E.; de Carvalho, J.F.; Zohner, C.M.; Crowther, T.W. The relative roles of local climate adaptation and phylogeny in determining leaf-out timing of temperate tree species. *For. Ecosytems* **2017**, *4*, 2–7. [CrossRef]

27. Jastrzębowski, S.; Ukalska, J. Dynamics of epicotyl emergence of *Quercus robur* from different climatic regions is strongly driven by post-germination temperature and humidity conditions. *Dendrobiology* **2019**, *81*, 73–85. [CrossRef]

28. Cooper, H.F.; Grady, K.C.; Cowan, J.A.; Best, R.J.; Allan, G.J.; Whitham, T.G. Genotypic variation in phenological plasticity: Reciprocal common gardens reveal adaptive responses to warmer springs but not to fall frost. *Glob. Chang. Biol.* **2019**, *25*, 187–200. [CrossRef] [PubMed]

29. Fang, X.-W.; Zhang, J.-J.; Xu, D.-H.; Pang, J.; Gao, T.-P.; Zhang, C.-H.; Li, F.-M.; Turner, N.C. Seed germination of Caragana species from different regions is strongly driven by environmental cues and not phylogenetic signals. *Sci. Rep.* **2017**, *7*, 11248. [CrossRef] [PubMed]

30. Daws, M.I.; Cleland, H.; Chmielarz, P.; Gorian, F.; Leprince, O.; Mullins, C.E.; Thanos, C.A.; Vandvik, V.; Pritchard, H.W. Variable desiccation tolerance in *Acer pseudoplatanus* seeds in relation to developmental conditions: A case of phenotypic recalcitrance? *Funct. Plant. Boil.* **2006**, *33*, 59–66. [CrossRef]

31. Moles, A.T.; Ackerly, D.D.; Webb, C.O.; Tweddle, J.C.; Dickie, J.B.; Pitman, A.J.; Westoby, M. Factors that shape seed mass evolution. *Proc. Natl. Acad. Sci. USA* **2005**, *102*, 10540–10544. [CrossRef]

32. Arène, F.; Affre, L.; Doxa, A.; Saatkamp, A. Temperature but not moisture response of germination shows phylogenetic constraints while both interact with seed mass and lifespan. *Seed Sci. Res.* **2017**, *27*, 110–120. [CrossRef]

33. Karley, A.J.; Leigh, R.A.; Sanders, D. Where do all the ions go? The cellular basis of differential ion accumulation in leaf cells. *Trends Plant. Sci.* **2000**, *5*, 465–470. [CrossRef]

34. Hayat, S.; Hayat, Q.; Alyemeni, M.N.; Wani, A.S.; Pichtel, J.; Ahmad, A. Role of proline under changing environments: A review. *Plant Signal Behav.* **2012**, *7*, 1456–1466. [CrossRef]

35. Anjum, N.A.; Sofo, A.; Scopa, A.; Roychoudhury, A.; Gill, S.S.; Iqbal, M.; Lukatkin, A.S.; Pereira, E.; Duarte, A.C.; Ahmad, I. Lipids and proteins–major targets of oxidative modifications in abiotic stressed plants. *Environ. Sci. Pollut. Res Int.* **2015**, *22*, 4099–4121. [CrossRef] [PubMed]

36. Sharma, A.; Shahzad, B.; Kumar, V.; Kohli, S.K.; Sidhu, G.P.S.; Bali, A.S.; Handa, N.; Kapoor, D.; Bhardwaj, R.; Zheng, B. Phytohormones Regulate Accumulation of Osmolytes Under Abiotic Stress. *Biomolecules* **2019**, *9*, 285. [CrossRef]

37. Marriboina, S.; Attipalli, R.R. Hydrophobic cell-wall barriers and vacuolar sequestration of Na+ ions are among the key mechanisms conferring high salinity tolerance in a biofuel tree species, *Pongamia pinnata* L. pierre. *Environ. Exp. Bot.* **2020**, *171*, 103949. [CrossRef]

38. Reddy, A.R.; Chaitanya, K.V.; Vivekanandan, M. Drought-induced responses of photosynthesis and antioxidant metabolism in higher plants. *J. Plant Physiol.* **2004**, *161*, 1189–1202. [CrossRef] [PubMed]

39. Xiong, L.; Wang, R.-G.; Mao, G.; Koczan, J.M. Identification of Drought Tolerance Determinants by Genetic Analysis of Root Response to Drought Stress and Abscisic Acid1. *Plant Physiol.* **2006**, *142*, 1065–1074. [CrossRef] [PubMed]

40. Zhang, J.; Jia, W.; Yang, J.; Ismail, A.M. Role of ABA in integrating plant responses to drought and salt stresses. *Field Crop. Res.* **2006**, *97*, 111–119. [CrossRef]

41. Sah, S.K.; Reddy, K.R.; Li, J. Abscisic Acid and Abiotic Stress Tolerance in Crop Plants. *Front. Plant Sci.* **2016**, *7*, 63. [CrossRef] [PubMed]

42. Nirala, D.; Pant, N.; Rawat, M. Response of tree roots to drought condition: A review. *Int. J. Chem.* **2019**, *7*, 824–826. [CrossRef]

43. Joslin, J.D.; Wolfe, M.H.; Hanson, P.J. Effects of altered water regimes on forest root systems. *New Phytol.* **2000**, *147*, 117–129. [CrossRef]

44. Iqbal, N.; Khan, N.A.; Ferrante, A.; Trivellini, A.; Francini, A.; Khan, M.I.R. Ethylene Role in Plant Growth, Development and Senescence: Interaction with Other Phytohormones. *Front. Plant Sci.* **2017**, *8*, 6484. [CrossRef]

45. Khan, N.A.; Khan, M.I.R.; Ferrante, A.; Poor, P. Editorial: Ethylene: A Key Regulatory Molecule in Plants. *Front. Plant Sci.* **2017**, *8*, 1782. [CrossRef] [PubMed]

46. Rust, S.; Roloff, A. Reduced photosynthesis in old oak (*Quercus robur*): The impact of crown and hydraulic architecture. *Tree Physiol.* **2002**, *22*, 597–601. [CrossRef] [PubMed]

47. Niinemets, Ü. Controls on the emission of plant volatiles through stomata: A sensitivity analysis. *J. Geophys. Res. Space Phys.* **2003**, *108*. [CrossRef]

48. Tombesi, S.; Nardini, A.; Frioni, T.; Soccolini, M.; Zadra, C.; Farinelli, D.; Poni, S.; Palliotti, A. Stomatal closure is induced by hydraulic signals and maintained by ABA in drought-stressed grapevine. *Sci. Rep.* **2015**, *5*, 12449. [CrossRef]

49. Agurla, S.; Gahir, S.; Munemasa, S.; Murata, Y.; Raghavendra, A.S. Mechanism of Stomatal Closure in Plants Exposed to Drought and Cold Stress. *Kidney Dev. Dis.* **2018**, 215–232. [CrossRef]

50. Leuschner, C.; Wedde, P.; Lübbe, T. The relation between pressure–volume curve traits and stomatal regulation of water potential in five temperate broadleaf tree species. *Ann. For. Sci.* **2019**, *76*, 60. [CrossRef]

51. Le Provost, G.; Domergue, F.; Lalanne, C.; Campos, P.R.; Grosbois, A.; Bert, D.; Meredieu, C.; Danjon, F.; Plomion, C.; Gion, J.M. Soil water stress affects both cuticular wax content and cuticle-related gene expression in young saplings of maritime pine (*Pinus pinaster* Ait). *BMC Plant Biol.* **2013**, *13*, 95. [CrossRef]

52. Griñán, I.; Rodríguez, P.; Nouri, H.; Wang, R.; Huang, G.; Morales, D.; Corell, M.; Pérez-López, D.; Centeno, A.; Martin-Palomo, M.; et al. Leaf mechanisms involved in the response of *Cydonia oblonga* trees to water stress and recovery. *Agric. Water Manag.* **2019**, *221*, 66–72. [CrossRef]

53. Abdulrahman, A.A.; Oladele, F.A. Leaf size and transpiration rates in Agave americana and Aloe vera. *Phytol. Balcan.* **2017**, *23*, 95–100. [CrossRef]

54. Lazaridou, M.; Koutroubas, S.D. Drought effect on water use efficiency of berseem clover at various growth stages. In *New Directions for a Diverse Planet: Proceedings of the 4th International Crop Science Congress Brisbane*; Australian Society of Agronomy Inc.: Melbourne, Australia, 2004; Volume 26, ISBN 1920842225.

55. Sanchez, A.C.; Haq, N.; Assogbadjo, A.E. Variation in baobab (*Adansonia digitata* L.) leaf morphology and its relation to drought tolerance. *Genet. Resour Crop Evol.* **2010**, *57*, 17–25. [CrossRef]

56. Flexas, J.; Medrano, H. Drought-inhibition of Photosynthesis in C3 Plants: Stomatal and Non-stomatal Limitations Revisited. *Ann. Bot.* **2002**, *89*, 183–189. [CrossRef]

57. Van Lun, M.; Hub, J.S.; Van Der Spoel, D.; Andersson, I. CO_2 and O_2 Distribution in Rubisco Suggests the Small Subunit Functions as a CO_2 Reservoir. *J. Am. Chem. Soc.* **2014**, *136*, 3165–3171. [CrossRef]

58. Rennenberg, H.; Loreto, F.; Polle, A.; Brilli, F.; Fares, S.; Beniwal, R.S.; Gessler, A. Physiological Responses of Forest Trees to Heat and Drought. *Plant Boil.* **2006**, *8*, 556–571. [CrossRef] [PubMed]

59. Gill, S.S.; Tuteja, N. Reactive oxygen species and antioxidant machinery in abiotic stress tolerance in crop plants. *Plant Physiol. Biochem.* **2010**, *48*, 909–930. [CrossRef] [PubMed]

60. Tikkanen, M.; Aro, E.-M. Thylakoid protein phosphorylation in dynamic regulation of photosystem II in higher plants. *Bioenergetics* **2012**, *1817*, 232–238. [CrossRef] [PubMed]

61. Haldimann, P.; Feller, U. Inhibition of photosynthesis by high temperature in oak (*Quercus pubescens* L.) leaves grown under natural conditions closely correlates with a reversible heat-dependent reduction of the activation state of ribulose-1,5-bisphosphate carboxylase/oxygenase. *Plant Cell Environ.* **2004**, *27*, 1169–1183. [CrossRef]

62. Johnson, M.P. Photosynthesis. *Essays Biochem.* **2016**, *31*, 255–273. [CrossRef]

63. Foyer, C.H. Reactive oxygen species, oxidative signaling and the regulation of photosynthesis. *Environ. Exp. Bot.* **2018**, *154*, 134–142. [CrossRef]

64. Hideg, Éva A comparative study of fluorescent singlet oxygen probes in plant leaves. *Open Life Sci.* **2008**, *3*, 273–284. [CrossRef]

65. Dietz, K.-J.; Turkan, I.; Krieger-Liszkay, A. Redox- and Reactive Oxygen Species-Dependent Signaling into and out of the Photosynthesizing Chloroplast. *Plant Physiol.* **2016**, *171*, 1541–1550. [CrossRef]

66. Sweetlove, L.J.; Lytovchenko, A.; Morgan, M.; Nunes-Nesi, A.; Taylor, N.L.; Baxter, C.J.; Eickmeier, I.; Fernie, A.R. Mitochondrial uncoupling protein is required for efficient photosynthesis. *Proc. Natl. Acad. Sci. USA* **2006**, *103*, 19587–19592. [CrossRef] [PubMed]

67. Noctor, G.; De Paepe, R.; Foyer, C.H. Mitochondrial redox biology and homeostasis in plants. *Trends Plant Sci.* **2007**, *12*, 125–134. [CrossRef] [PubMed]

68. Schwarzländer, M.; Finkemeier, I. Mitochondrial energy and redox signaling in plants. *Antioxid. Redox Signal.* **2013**, *18*, 2122–2144. [CrossRef] [PubMed]

69. Han, D.; Williams, E.; Cadenas, E. Mitochondrial respiratory chain-dependent generation of superoxide anion and its release into the intermembrane space. *Biochem. J.* **2001**, *353*, 411. [CrossRef] [PubMed]

70. Sharma, P.; Jha, A.B.; Dubey, R.S.; Pessarakli, M. Reactive Oxygen Species, Oxidative Damage, and Antioxidative Defense Mechanism in Plants under Stressful Conditions. *J. Bot.* **2012**, *2012*, 1–26. [CrossRef]

71. Pospíšil, P. Production of Reactive Oxygen Species by Photosystem II as a Response to Light and Temperature Stress. *Front. Plant Sci.* **2016**, *7*, 1950. [CrossRef]

72. Mittler, R.; Vanderauwera, S.; Suzuki, N.; Miller, G.; Tognetti, V.B.; Vandepoele, K.; Gollery, M.; Shulaev, V.; Van Breusegem, F. ROS signaling: The new wave? *Trends Plant Sci.* **2011**, *16*, 300–309. [CrossRef] [PubMed]

73. Feng, H.; Sun, K.; Li, M.; Li, H.; Li, X.; Li, Y.; Wang, Y. The expression, function and regulation of mitochondrial alternative oxidase under biotic stresses. *Mol. Plant Pathol.* **2010**, *11*, 429–440. [CrossRef]

74. Vanlerberghe, G.C. Alternative Oxidase: A Mitochondrial Respiratory Pathway to Maintain Metabolic and Signaling Homeostasis during Abiotic and Biotic Stress in Plants. *Int. J. Mol. Sci.* **2013**, *14*, 6805–6847. [CrossRef]

75. Saha, B.; Borovskii, G.; Panda, S.K. Alternative oxidase and plant stress tolerance. *Plant Signal. Behav.* **2016**, *11*, e1256530. [CrossRef]

76. Brosché, M.; Overmyer, K.; Wrzaczek, M.; Kangasjärvi, J.; Kangasjärvi, S. Stress signaling III: Reactive oxygen species (ROS). In *Abiotic Stress Adaptation in Plants*; Pareek, A., Sopory, S., Bohnert, H., Eds.; Springer: Dordrecht, The Netherlands, 2009; pp. 91–102. ISBN 978-90-481-3112-9.

77. Źróbek-Sokolnik, A. Temperature stress and responses of plants. In *Environmental Adaptations and Stress Tolerance of Plants in the Era of Climate Change*; Ahmad, P., Prasad, M.N.V., Eds.; Springer: New York, NY, USA, 2012; pp. 113–134. ISBN 978-1-4614-0814-7.

78. Ruelland, E.; Zachowski, A. How plants sense temperature. *Environ. Exp. Bot.* **2010**, *69*, 225–232. [CrossRef]

79. Wahid, A.; Gelani, S.; Ashraf, M.; Foolad, M. Heat tolerance in plants: An overview. *Environ. Exp. Bot.* **2007**, *61*, 199–223. [CrossRef]

80. Kitao, M.; Lei, T.T.; Koike, T.; Tobita, H.; Maruyama, Y.; Matsumoto, Y.; Ang, L.-H. Temperature response and photoinhibition investigated by chlorophyll fluorescence measurements for four distinct species of dipterocarp trees. *Physiol. Plant* **2000**, *109*, 284–290. [CrossRef]

81. Mittler, R. Oxidative stress, antioxidants and stress tolerance. *Trends Plant Sci.* **2002**, *7*, 405–410. [CrossRef]

82. Suzuki, N.; Miller, G.; Salazar, C.; Mondal, H.A.; Shulaev, E.; Cortes, D.F.; Shuman, J.L.; Luo, X.; Shah, J.; Schlauch, K.; et al. Temporal-spatial interaction between reactive oxygen species and abscisic acid regulates rapid systemic acclimation in plants. *Plant Cell* **2013**, *25*, 3553–3569. [CrossRef]

83. Pukacka, S.; Malec, M.; Ratajczak, E. ROS production and antioxidative system activity in embryonic axes of *Quercus robur* seeds under different desiccation rate conditions. *Acta Physiol. Plant.* **2011**, *33*, 2219–2227. [CrossRef]

84. Pukacka, S.; Ratajczak, E. Production and scavenging of reactive oxygen species in *Fagus sylvatica* seeds during storage at varied temperature and humidity. *J. Plant Physiol.* **2005**, *162*, 873–885. [CrossRef]

85. Choudhury, S.; Panda, P.; Sahoo, L.; Panda, S.K. Reactive oxygen species signaling in plants under abiotic stress. *Plant Signal. Behav.* **2013**, *8*, e23681. [CrossRef]

86. Suzuki, N.; Koussevitzky, S.; Mittler, R.; Miller, G. ROS and redox signalling in the response of plants to abiotic stress. *Plant Cell Environ.* **2012**, *35*, 259–270. [CrossRef]

87. Kalemba, E.M.; Suszka, J.; Ratajczak, E. The role of oxidative stress in determining the level of viability of black poplar (*Populus nigra*) seeds stored at different temperatures. *Funct. Plant. Boil.* **2015**, *42*, 630–642. [CrossRef]

88. Ratajczak, E.; Małecka, A.; Bagniewska-Zadworna, A.; Kalemba, E.M. The production, localization and spreading of reactive oxygen species contributes to the low vitality of long-term stored common beech (*Fagus sylvatica* L.) seeds. *J. Plant Physiol.* **2015**, *174*, 147–156. [CrossRef] [PubMed]

89. Ratajczak, E.; Staszak, A.M.; Wojciechowska, N.; Bagniewska-Zadworna, A.; Dietz, K.J. Regulation of thiol metabolism as a factor that influences the development and storage capacity of beech seeds. *J. Plant Physiol.* **2019**, *239*, 61–70. [CrossRef] [PubMed]

90. Suzuki, N.; Mittler, R. Reactive oxygen species and temperature stresses: A delicate balance between signaling and destruction. *Physiol. Plant* **2006**, *126*, 45–51. [CrossRef]

91. Repetto, M.; Semprine, J.; Boveris, A. Lipid Peroxidation: Chemical Mechanism, Biological Implications and Analytical Determination. *Lipid Peroxidation* **2012**, *1*, 3–30.

92. Leprince, O.; Buitink, J.; Hoekstra, F.A. Axes and cotyledons of recalcitrant seeds of *Castanea sativa* Mill. exhibit contrasting responses of respiration to drying in relation to desiccation sensitivity. *J. Exp. Bot.* **1999**, *50*, 1515–1524. [CrossRef]

93. Chmielowska-Bäk, J.; Izbiaå-Ska, K.; Deckert, J.; Izbiańska, K. Products of lipid, protein and RNA oxidation as signals and regulators of gene expression in plants. *Front. Plant Sci.* **2015**, *6*, 405. [CrossRef]

94. Bailly, C.; El-Maarouf-Bouteau, H.; Corbineau, F. From intracellular signaling networks to cell death: The dual role of reactive oxygen species in seed physiology. *Comptes Rendus Boil.* **2008**, *331*, 806–814. [CrossRef]

95. Berwal, M.K.; Ram, C.H. Superoxide Dismutase: A Stable Biochemical Marker for Abiotic Stress Tolerance in Higher Plants. In *Superoxide Dismutase*; Intech Open: London, UK, 2018. [CrossRef]

96. Caverzan, A.; Casassola, A.; Brammer, S.P. Reactive Oxygen Species and Antioxidant Enzymes Involved in Plant Tolerance to Stress. In *Abiotic and Biotic Stress in Plants*; Intech Open: London, UK, 2016; pp. 463–480. ISBN 978-953-51-2250-0.

97. Ratajczak, E.; Pukacka, S. Decrease in beech (*Fagus sylvatica*) seed viability caused by temperature and humidity conditions as related to membrane damage and lipid composition. *Acta Physiol. Plant* **2005**, *27*, 3–12. [CrossRef]

98. Nievola, C.C.; Carvalho, C.P.; Carvalho, V.; Rodrigues, E. Rapid responses of plants to temperature changes. *Temperature* **2017**, *4*, 371–405. [CrossRef]

99. Jactel, H.; Petit, J.; Desprez-Loustau, M.L.; Delzon, S.; Piou, D.; Battisti, A.; Koricheva, J. Drought effects on damage by forest insects and pathogens: A meta-analysis. *Glob. Change Biol.* **2012**, *18*, 267–276. [CrossRef]

100. Bentz, B.J.; Jönsson, A.M. Modeling Bark Beetle Responses to Climate Change. In *Bark Beetles*; Elsevier BV: Amsterdam, The Nederlands, 2015; pp. 533–553. ISBN 978-0-12-417156-5.

101. Bentz, B.J. Mountain pine beetle population sampling: Inferences from Lindgren pheromone traps and tree emergence cages. *Can. J. For. Res.* **2006**, *36*, 351–360. [CrossRef]

102. Keeling, C.I.; Yuen, M.M.S.; Liao, N.Y.; Docking, T.R.; Chan, S.K.; Taylor, G.A.; Palmquist, D.L.; Jackman, S.D.; Nguyen, A.; Li, M.; et al. Draft genome of the mountain pine beetle, Dendroctonus ponderosae Hopkins, a major forest pest. *Genome Boil.* **2013**, *14*, R27. [CrossRef]

103. Marini, L.; Ayres, M.P.; Battisti, A.; Faccoli, M. Climate affects severity and altitudinal distribution of outbreaks in an eruptive bark beetle. *Clim. Chang.* **2012**, *115*, 327–341. [CrossRef]

104. Raffa, K.F.; Aukema, B.H.; Bentz, B.J.; Carroll, A.L.; Hicke, J.A.; Kolb, T.E.; Björkman, C.; Niemelä, P. Responses of tree-killing bark beetles to a changing climate. In *Climate Change and Insect Pests*; CABI: Warszawa, Poland, 2015; pp. 173–201. ISBN 9781780643786.

105. Starzyk, J.R. Species characteristics (Charakterystyka gatunku). In *The bark beetle and its role in forest ecosystems (Kornik drukarz i jego rola w ekosystemach leśnych)*; Grodzki, W., Ed.; CILP: Warszawa, Poland, 2013; pp. 99–103. ISBN 978-83-63895-08-2.

106. Brzeziecki, B.; Hilszczański, J.; Kowalski, T.; Łakomy, P.; Małek, S.; Miścicki, S.; Modrzyński, J.; Sowa, J.; Starzyk, J.R. The problem of mortality of spruce stands in the "Puszcza Białowieska" Leśny Kompleks Promocyjny (Problem masowego zamierania drzewostanów świerkowych w Leśnym Kompleksie Promocyjnym "Puszcza Białowieska"). *Sylwan* **2018**, 162.

107. Starzyk, J.R. Abiotic and anthropogenic factors (Czynniki abiotyczne i antropogeniczne). In *The Bark Beetle and its Role in Forest Ecosystems (Kornik drukarz i jego rola w ekosystemach leśnych)*; Grodzki, W., Ed.; CILP: Warszawa, Poland, 2013; pp. 99–103. ISBN 978-83-63895-08-2.

108. Virtanen, T.; Neuvonen, S.; Nikula, A.; Varama, M.; Niemelä, P. Climate change and the risks of Neodiprion sertifer outbreaks on Scots pine. *Silva. Fenn.* **1996**, *30*, 169–177. [CrossRef]

109. Ramsfield, T.; Bentz, B.; Faccoli, M.; Jactel, H.; Brockerhoff, E.G. Forest health in a changing world: Effects of globalization and climate change on forest insect and pathogen impacts. *Forests* **2016**, *89*, 245–252. [CrossRef]

110. Jaworski, J.; Hilszczański, T. Impact of temperature and humidity changes on development cycles and the importance of insects in forest ecosystems in connection with probable climate changes. (Wpływ zmian temperatury i wilgotności na cykle rozwojowe i znaczenie owadów w ekosystemach leśnych w związku z prawdopodobnymi zmianami klimatycznymi). *For. Res. Pap.* **2013**, *74*, 345–355. [CrossRef]

111. Hódar, J.A.; Zamora, R. Herbivory and climatic warming: A Mediterranean outbreaking caterpillar attacks a relict, boreal pine species. *Biodivers. Conserv.* **2004**, *13*, 493–500. [CrossRef]

112. Dickie, I.A.; Bufford, J.L.; Cobb, R.C.; Desprez-Loustau, M.-L.; Grelet, G.; Hulme, P.E.; Klironomos, J.; Makiola, A.; Nuñez, M.A.; Pringle, A.; et al. The emerging science of linked plant-fungal invasions. *New Phytol.* **2017**, *215*, 1314–1332. [CrossRef]

113. Britton, K.O.; Liebhold, A.M. One world, many pathogens! *New Phytol.* **2013**, *197*, 9–10. [CrossRef]

114. Lonsdale, D. Review of oak mildew, with particular reference to mature and veteran trees in Britain. *Arboric. J.* **2015**, *37*, 61–84. [CrossRef]

115. Guo, C.; Xu, J.; Yang, L.; Guo, X.; Liao, J.; Zheng, X.; Zhang, Z.; Chen, X.; Yang, K.; Wang, M. Life cycle evaluation of greenhouse gas emissions of a highway tunnel: A case study in China. *J. Clean. Prod.* **2019**, *211*, 972–980. [CrossRef]

116. Marçais, B.; Desprez-Loustau, M.L. European oak powdery mildew: Impact on trees, effects of environmental factors, and potential effects of climate change. *Ann. For. Sci.* **2014**, *71*, 633–642. [CrossRef]

117. Jarvis, W.R.; Gubler, W.D.; Grove, G.G. Epidemiology of powdery mildews in agricultural pathosystems. In *The Powdery Mildews, A Comprehensive Treatise*; Bélanger, R.R., Bushnell, W.R., Dik, A.J., Carver, T.L.W., Eds.; APS Press: St. Paul, MI, USA, 2002; pp. 169–199.

118. Asher, M.J.C.; Williams, G.E. Forecasting the national incidence of sugar-beet powdery mildew from weather data in Britain. *Plant Pathol.* **1991**, *40*, 100–107. [CrossRef]

119. Mmbaga, M.T. Ascocarp formation and survival and primary inoculum production in Erysiphe (sect. Microsphaera) pulchra in dogwood powdery mildew. *Ann. Appl. Boil.* **2002**, *141*, 153–161. [CrossRef]

120. Penczykowski, R.M.; Walker, E.; Soubeyrand, S.; Laine, A.L. Linking winter conditions to regional disease dynamics in a wild plant–pathogen metapopulation. *New Phytol.* **2015**, *205*, 1142–1152. [CrossRef]

121. Camenen, E.; Porté, A.J.; Garzón, M.B. American trees shift their niches when invading Western Europe: Evaluating invasion risks in a changing climate. *Ecol. Evol.* **2016**, *6*, 7263–7275. [CrossRef]

122. Johnsen, O.; Daehlen, O.G.; Østreng, G.; Skrøppa, T.; Dæhlen, O.G. Daylength and temperature during seed production interactively affect adaptive performance of Picea abies progenies. *New Phytol.* **2005**, *168*, 589–596. [CrossRef] [PubMed]

123. Caignard, T.; Kremer, A.; Firmat, C.; Nicolas, M.; Venner, S.; Delzon, S. Increasing spring temperatures favor oak seed production in temperate areas. *Sci. Rep.* **2017**, *7*, 8555. [CrossRef]

124. Rossi, S. Local adaptations and climate change: Converging sensitivity of bud break in black spruce provenances. *Int. J. Biometeorol.* **2015**, *59*, 827–835. [CrossRef]

125. Ford, K.R.; Harrington, C.A.; Bansal, S.; Gould, P.J.; Clair, J.B.S. Will changes in phenology track climate change? A study of growth initiation timing in coast Douglas-fir. *Glob. Chang. Boil.* **2016**, *22*, 3712–3723. [CrossRef] [PubMed]

126. Lankau, R.A.; Zhu, K.; Ordonez, A. Mycorrhizal strategies of tree species correlate with trailing range edge responses to current and past climate change. *Ecology* **2015**, *96*, 1451–1458. [CrossRef]

127. Lau, J.A.; Lennon, J.T.; Heath, K.D. Trees harness the power of microbes to survive climate change. *Proc. Natl. Acad. Sci. USA* **2017**, *114*, 11009–11011. [CrossRef] [PubMed]

128. Cavin, L.; Jump, A.S. Highest drought sensitivity and lowest resistance to growth suppression are found in the range core of the tree *Fagus sylvatica* L. not the equatorial range edge. *Glob. Chang. Boil.* **2016**, *23*, 362–379. [CrossRef]

129. Savi, T.; Bertuzzi, S.; Branca, S.; Tretiach, M.; Nardini, A. Drought-induced xylem cavitation and hydraulic detorientation: Risk factors for urban trees under climate change? *New Phytol.* **2015**, *205*, 1106–1116. [CrossRef]

130. Pretzsch, H.; Biber, P.; Uhl, E.; Dahlhausen, J.; Schütze, G.; Perkins, D.; Rötzer, T.; Caldentey, J.; Koike, T.; Van Con, T.; et al. Climate change accelerates growth of urban trees in metropolises worldwide. *Sci. Rep.* **2017**, *7*, 15403. [CrossRef]

131. Medlyn, B.E.; Dreyer, E.; Ellsworth, D.; Forstreuter, M.; Harley, P.C.; Kirschbaum, M.U.F.; Le Roux, X.; Montpied, P.; Strassemeyer, J.; Walcroft, A.; et al. Temperature response of parameters of a biochemically based model of photosynthesis. II. A review of experimental data. *Plant Cell Environ.* **2002**, *25*, 253–264. [CrossRef]

132. De Kort, H.; Vander Mijnsbrugge, K.; Vandepitte, K.; Mergeay, J.; Ovaskainen, O.; Honnay, O. Evolution, plasticity and evolving plasticity of phenology in the tree species *Alnus glutinosa*. *J. Evol. Biol.* **2016**, *29*, 253–264. [CrossRef]

133. Jump, A.S.; Peñuelas, J. Running to stand still: Adaptation and the response of plants to rapid climate change. *Ecol. Lett.* **2005**, *8*, 1010–1020. [CrossRef]

134. Williams, B.L.; Brawn, J.D.; Paige, K.N. Landscape scale genetic effects of habitat fragmentation on a high gene flow species: *Speyeria idalia* (Nymphalidae). *Mol. Ecol.* **2003**, *12*, 11–20. [CrossRef]

135. Gutschick, V.P.; BassiriRad, H. Extreme events as shaping physiology, ecology, and evolution of plants: Toward a unified definition and evaluation of their consequences. *New Phytol.* **2003**, *160*, 21–42. [CrossRef]

136. DeWitt, T.J.; Sih, A.; Wilson, D.S. Costs and limits of phenotypic plasticity. *Trends Ecol. Evol.* **1998**, *13*, 77–81. [CrossRef]

137. Peñuelas, J.; Lloret, F.; Montoya, R. Severe drought effects on Mediterranean woody flora in Spain. *For. Sci.* **2001**, *47*, 214–218. [CrossRef]

138. Peñuelas, J.; Boada, M. A global change-induced biome shift in the Montseny mountains (NE Spain). *Glob. Chang. Boil.* **2003**, *9*, 131–140. [CrossRef]

139. Walther, G.R. Plants in a warmer world. In *Perspectives in Plant Ecology, Evolution and Systematics*; Edwards, P., Ed.; Stockton Press: Jena, Germany, 2003; Volume 6, pp. 169–185. ISBN 14338319.

140. Pliūra, A.; Jankauskienė, J.; Bajerkevičienė, G.; Lygis, V.; Suchockas, V.; Labokas, J.; Verbylaitė, R. Response of juveniles of seven forest tree species and their populations to different combinations of simulated climate change-related stressors: Spring-frost, heat, drought, increased UV radiation and ozone concentration under elevated CO_2 level. *J. Plant Res.* **2019**, *132*, 789–811. [CrossRef]

141. De Jong, G. Evolution of phenotypic plasticity: Patterns of plasticity and the emergence of ecotypes. *New Phytol.* **2005**, *166*, 101–118. [CrossRef]

142. Dobrowolska, D. Vitality of European Beech (*Fagus sylvatica* L.) at the Limit of Its Natural Range in Poland. *Pol. J. Ecol.* **2015**, *63*, 260–272. [CrossRef]

143. Arana, M.V.; Gonzalez-Polo, M.; Martinez-Meier, A.; Gallo, L.A.; Benech-Arnold, R.L.; Sánchez, R.A.; Batlla, D. Seed dormancy responses to temperature relate to Nothofagus species distribution and determine temporal patterns of germination across altitudes in Patagonia. *New Phytol.* **2016**, *209*, 507–520. [CrossRef]

144. Walck, J.L.; Hidayati, S.N.; Dixon, K.W.; Thompson, K.; Poschlod, P. Climate change and plant regeneration from seed. *Glob. Chang. Boil.* **2011**, *17*, 2145–2161. [CrossRef]

145. O'Sullivan, O.S.; Heskel, M.A.; Reich, P.B.; Tjoelker, M.G.; Weerasinghe, L.K.; Penillard, A.; Zhu, L.; Egerton, J.J.G.; Bloomfield, K.J.; Creek, D.; et al. Thermal limits of leaf metabolism across biomes. *Glob. Chang. Boil.* **2016**, *23*, 209–223. [CrossRef]

146. Baskin, J.M.; Baskin, C.C. Seed germination ecophysiology of jeffersonia diphylla, a perennial herb of mesic deciduous forests. *Am. J. Bot.* **1989**, *76*, 1073–1080. [CrossRef]

147. Jansen, P.A.; Bongers, F.; Hemerik, L. Seed mass and mast seeding enhance dispersal by a neotropical scatter-hoarding rodent. *Ecol. Monogr.* **2004**, *74*, 569–589. [CrossRef]

148. Gómez, J.M. Bigger is not always better: Conflicting selective pressures on seed size in quercus ILEX. *Evolution* **2004**, *58*, 71–80. [CrossRef] [PubMed]

149. Davis, C.C.; Willis, C.G.; Primack, R.B.; Miller-Rushing, A.J. The importance of phylogeny to the study of phenological response to global climate change. *Philos. Trans. R. Soc. B Boil. Sci.* **2010**, *365*, 3201–3213. [CrossRef]

150. Carta, A.; Hanson, S.; Müller, J.V. Plant regeneration from seeds responds to phylogenetic relatedness and local adaptation in Mediterranean Romulea (Iridaceae) species. *Ecol. Evol.* **2016**, *6*, 4166–4178. [CrossRef]

151. Willis, C.G.; Ruhfel, B.; Primack, R.B.; Miller-Rushing, A.J.; Davis, C.C. Phylogenetic patterns of species loss in Thoreau's woods are driven by climate change. *Proc. Natl. Acad. Sci. USA* **2008**, *105*, 17029–17033. [CrossRef]

152. Davies, T.J.; Wolkovich, E.M.; Kraft, N.J.B.; Salamin, N.; Allen, J.M.; Ault, T.R.; Betancourt, J.L.; Bolmgren, K.; Cleland, E.E.; Cook, B.I.; et al. Phylogenetic conservatism in plant phenology. *J. Ecol.* **2013**, *101*, 1520–1530. [CrossRef]

153. Richardson, B.A.; Chaney, L.; Shaw, N.L.; Still, S.M. Will phenotypic plasticity affecting flowering phenology keep pace with climate change? *Glob. Change Biol.* **2017**, *23*, 2499–2508. [CrossRef]

154. Levin, N.A. Flowering Phenology in Relation to Adaptive Radiation. *Syst. Bot.* **2006**, *31*, 239–246. [CrossRef]

155. De Casas, R.R.; Kovach, K.; Dittmar, E.; Barua, D.; Barco, B.; Donohue, K. Seed after-ripening and dormancy determine adult life history independently of germination timing. *New Phytol.* **2012**, *194*, 868–879. [CrossRef]

156. Elwell, A.L.; Gronwall, D.S.; Miller, N.D.; Spalding, E.P.; Durham Brooks, T.L. Separating parental environment from seed size effects on next generation growth and development in Arabidopsis. *Plant Cell Environ.* **2011**, *34*, 291–301. [CrossRef]

157. Billington, H.L.; Pelham, J. Genetic Variation in the Date of Budburst in Scottish Birch Populations: Implications for Climate Change. *Funct. Ecol.* **1991**, *5*, 403. [CrossRef]

158. Dahlquist, R.M.; Prather, T.S.; Stapleton, J.J. Time and Temperature Requirements for Weed Seed Thermal Death. *Weed Sci.* **2007**, *55*, 619–625. [CrossRef]

159. Kim, D.H.; Han, S.H. Direct Effects on Seed Germination of 17 Tree Species under Elevated Temperature and CO_2 Conditions. *Open Life Sci.* **2018**, *13*, 137–148. [CrossRef]

160. Peterson, M.E.; Daniel, R.M.; Danson, M.J.; Eisenthal, R. The dependence of enzyme activity on temperature: Determination and validation of parameters. *Biochem. J.* **2007**, *402*, 331–337. [CrossRef]

161. De Villalobos, A.; Pelaez, D.; Boo, R.; Mayor, M.; Elia, O. Effect of high temperatures on seed germination of *Prosopis caldenia* Burk. *J. Arid. Environ.* **2002**, *52*, 371–378. [CrossRef]

162. Harrington, C.A.; Gould, P.J.; St.Clair, J.B. Modeling the effects of winter environment on dormancy release of Douglas-fir. *For. Ecol. Manag.* **2010**, *259*, 798–808. [CrossRef]

163. Duputié, A.; Rutschmann, A.; Ronce, O.; Chuine, I. Phenological plasticity will not help all species adapt to climate change. *Glob. Chang. Boil.* **2015**, *21*, 3062–3073. [CrossRef]

164. Vegis, A. Dormancy in Higher Plants. *Annu. Rev. Plant Physiol.* **1964**, *15*, 185–224. [CrossRef]

165. Fu, Y.H.; Piao, S.; Vitasse, Y.; Zhao, H.; De Boeck, H.J.; Liu, Q.; Yang, H.; Weber, U.; Hänninen, H.; Janssens, I.A. Increased heat requirement for leaf flushing in temperate woody species over 1980–2012: Effects of chilling, precipitation and insolation. *Glob. Chang. Boil.* **2015**, *21*, 2687–2697. [CrossRef]

166. Hänninen, H.; Kramer, K. A framework for modelling the annual cycle of trees in boreal and temperate regions. *Silva. Fenn.* **2007**, *41*, 167–205. [CrossRef]

167. West-Eberhard, M.J. Toward a Modern Revival of Darwin's Theory of Evolutionary Novelty. *Philos. Sci.* **2008**, *75*, 899–908. [CrossRef]

168. Jump, A.S.; Marchant, R.; Peñuelas, J.; Marchant, R. Environmental change and the option value of genetic diversity. *Trends Plant Sci.* **2009**, *14*, 51–58. [CrossRef]

169. Nicotra, A.; Atkin, O.; Bonser, S.; Davidson, A.; Finnegan, E.J.; Mathesius, U.; Poot, P.; Purugganan, M.; Richards, C.; Valladares, F.; et al. Plant phenotypic plasticity in a changing climate. *Trends Plant Sci.* **2010**, *15*, 684–692. [CrossRef] [PubMed]

170. Lande, R. Adaptation to an extraordinary environment by evolution of phenotypic plasticity and genetic assimilation. *J. Evol. Boil.* **2009**, *22*, 1435–1446. [CrossRef] [PubMed]

171. Van Kleunen, M.; Fischer, M. Adaptive evolution of plastic foraging responses in a clonal plant. *Ecology* **2001**, *82*, 3309–3319. [CrossRef]

172. Van Kleunen, M.; Lenssen, J.P.M.; Fischer, M.; De Kroon, H. Selection on phenotypic plasticity of morphological traits in response to flooding and competition in the clonal shore plant Ranunculus reptans. *J. Evol. Boil.* **2007**, *20*, 2126–2137. [CrossRef] [PubMed]

173. Bräutigam, K.; Vining, K.J.; Lafon-Placette, C.; Fossdal, C.G.; Mirouze, M.; Marcos, J.G.; Fluch, S.; Fraga, M.F.; Guevara, M.; Ángeles, L.; et al. Epigenetic regulation of adaptive responses of forest tree species to the environment. *Ecol. Evol.* **2013**, *3*, 399–415. [CrossRef]

174. Zhang, Y.Y.; Fischer, M.; Colot, V.; Bossdorf, O. Epigenetic variation creates potential for evolution of plant phenotypic plasticity. *New Phytol.* **2013**, *197*, 314–322. [CrossRef]

175. Sow, M.D.; Segura, V.; Chamaillard, S.; Jorge, V.; Delaunay, A.; Lafon-Placette, C.; Fichot, R.; Faivre-Rampant, P.; Villar, M.; Brignolas, F.; et al. Narrow-sense heritability and PST estimates of DNA methylation in three Populus nigra L. populations under contrasting water availability. *Tree Genet. Genomes* **2018**, *14*, 78. [CrossRef]

176. Gourcilleau, D.; Bogeat-Triboulot, M.-B.; Thiec, D.; Lafon-Placette, C.; Delaunay, A.; Abu El-Soud, W.; Brignolas, F.; Maury, S. DNA methylation and histone acetylation: Genotypic variations in hybrid poplars, impact of water deficit and relationships with productivity. *Ann. For. Sci.* **2010**, *67*, 208. [CrossRef]

177. Le Gac, A.-L.; Lafon-Placette, C.; Chauveau, D.; Segura, V.; Delaunay, A.; Fichot, R.; Marron, N.; Le Jan, I.; Berthelot, A.; Bodineau, G.; et al. Winter-dormant shoot apical meristem in poplar trees shows environmental epigenetic memory. *J. Exp. Bot.* **2018**, *69*, 4821–4837. [CrossRef]

178. Raj, S.; Bräutigam, K.; Hamanishi, E.T.; Wilkins, O.; Thomas, B.R.; Schroeder, W.; Mansfield, S.D.; Plant, A.L.; Campbell, M.M. Clone history shapes Populus drought responses. *Proc. Natl. Acad. Sci. USA* **2011**, *108*, 12521–12526. [CrossRef] [PubMed]

179. Lafon-Placette, C.; Faivre-Rampant, P.; Delaunay, A.; Street, N.; Brignolas, F.; Maury, S. Methylome of DN ase I sensitive chromatin in Populus trichocarpa shoot apical meristematic cells: A simplified approach revealing characteristics of gene-body DNA methylation in open chromatin state. *New Phytol.* **2013**, *197*, 416–430. [CrossRef]

180. Bossdorf, O.; Richards, C.L.; Pigliucci, M. Epigenetics for ecologists. *Ecol Lett.* **2008**, *11*, 106–115. [CrossRef] [PubMed]

181. Sáez-Laguna, E.; Guevara, M.-A.; Díaz, L.-M.; Sánchez-Gómez, D.; Collada, C.; Aranda, I.; Cervera, M.-T. Epigenetic Variability in the Genetically Uniform Forest Tree Species Pinus pinea L. *PLoS ONE* **2014**, *9*, e103145. [CrossRef]

182. Skrøppa, T.; Johnsen, Ø. Patterns of adaptive genetic variation in forest tree species; the reproductive enviroment as an evolutionary force in Picea abies. *Agrofor. Sci. Policy Pract.* **2000**, *63*, 49–58.

183. Skrøppa, T.; Tollefsrud, M.M.; Sperisen, C.; Johnsen, Ø. Rapid change in adaptive performance from one generation to the next in Picea abies—Central European trees in a Nordic environment. *Tree Genet.Genomes* **2010**, *6*, 93–99. [CrossRef]

184. Ashcroft, M.B.; Gollan, J.R.; Batley, M. Combining citizen science, bioclimatic envelope models and observed habitat preferences to determine the distribution of an inconspicuous, recently detected introduced bee (Halictus smaragdulus Vachal Hymenoptera: Halictidae) in Australia. *Biol. Invasions* **2012**, *14*, 515–527. [CrossRef]

185. Deb, J.C.; Phinn, S.; Butt, N.; McAlpine, C.A. The impact of climate change on the distribution of two threatened Dipterocarp trees. *Ecol. Evol.* **2017**, *7*, 2238–2248. [CrossRef]

186. Hanewinkel, M.; Cullmann, D.A.; Schelhaas, M.-J.; Nabuurs, G.-J.; Zimmermann, N.E. Climate change may cause severe loss in the economic value of European forest land. *Nat. Clim. Chang.* **2013**, *3*, 203–207. [CrossRef]

187. Leech, S.M.; Almuedo, P.L.; O'Neill, G. Assisted migration: Adapting forest management to a changing climate. *J. Ecol. Manag.* **2011**, *12*, 18–34.

188. Dumroese, R.K.; Williams, M.I.; Stanturf, J.A.; Clair, J.B.S. Considerations for restoring temperate forests of tomorrow: Forest restoration, assisted migration, and bioengineering. *New For.* **2015**, *46*, 947–964. [CrossRef]

189. Erickson, V.; Aubry, C.; Berrang, P.; Blush, T.; Bower, A.; Crane, B.; Johnson, R. *Genetic Resource Management and Climate Change: Genetic Options for Adapting National Forests to Climate Change*; USDA Forest Service Forest Management: Washington, DC, USA, 2012.

190. Grady, K.C.; Kolb, T.E.; Ikeda, D.H.; Whitham, T.G. A bridge too far: Cold and pathogen constraints to assisted migration of riparian forests. *Restor. Ecol.* **2015**, *23*, 811–820. [CrossRef]

191. O'Neill, G.A.; Ukrainetz, N.K.; Carlson, M.R.; Cartwright, C.V.; Jaquish, B.C.; King, J.N.; Krakowski, J.; Russell, J.H.; Stoehr, M.; Xie, C.Y.; et al. *Assisted Migration to Address Climate Change in British Columbia: Recommendations for Interim Seed Transfer Standards*; BC Ministry of Forests and Range, Forest Sci Branch: Victoria, BC, Canada, 2008; Tech Rep. 048; ISBN 978-0-7726-6076-3.

192. St.Clair, J.B.; Howe, G.T. Strategies for conserving forest genetic resources in the face of climate change. *Turk. J. Bot.* **2011**, *35*, 403–409. [CrossRef]

193. Keppel, G.; Van Niel, K.P.; Wardell-Johnson, G.W.; Yates, C.J.; Byrne, M.; Mucina, L.; Schut, A.G.T.; Hopper, S.D.; Franklin, S.E. Refugia: Identifying and understanding safe havens for biodiversity under climate change. *Glob. Ecol. Biogeogr.* **2012**, *21*, 393–404. [CrossRef]

Article

Shifts in Climate–Growth Relationships of Sky Island Pines

Paula E. Marquardt [1,*], **Brian R. Miranda** [1] **and Frank W. Telewski** [2]

1   USDA Forest Service, Northern Research Station, 5985 Highway K, Rhinelander, WI 54501, USA;
    brian.r.miranda@usda.gov
2   W.J. Beal Botanical Garden, Department of Plant Biology, Michigan State University, East Lansing,
    MI 48824, USA; telewski@msu.edu
*   Correspondence: paula.marquardt@usda.gov

Received: 27 September 2019; Accepted: 30 October 2019; Published: 12 November 2019

Abstract: Rising temperatures and changes in precipitation may affect plant responses, and mountainous regions in particular are sensitive to the impacts of climate change. The Santa Catalina Mountains, near Tucson, Arizona, USA, are among the best known Madrean Sky Islands, which are defined by pine-oak forests. We compared the sensitivity and temporal stability of climate–growth relationships to quantify the growth responses of sympatric taxa of ponderosa pine to changing climate. Three taxa (three-needle, mixed-needle, and five-needle types) collected from southern slopes of two contact zones (Mt. Lemmon, Mt. Bigelow) were evaluated. Positive climate–growth correlations in these semiarid high-elevation pine forests indicated a seasonal shift from summer- to spring-dominant precipitation since 1950, which is a critical time for reproduction. Mixed- and five-needle types responded to winter precipitation, and growth was reduced for the five-needle type when spring conditions were dry. Growth trends in response to temperature and specific to site were observed, which indicated the climate signal can be weakened when data are combined into a single chronology. Significant fluctuations in temperature–growth correlations since 1950 occurred for all needle types. These results demonstrated a dramatic shift in sensitivity of annual tree growth to the seasonality of the limiting factor, and a climatic trend that increases local moisture stress may impact the stability of climate–growth relationships. Moreover, output from temperature–growth analyses based on ring-width data (for example from semiarid sites) that does not account for positive and negative growth trends may be adversely affected, potentially impacting climate reconstructions.

Keywords: dendrochronology; ecology; moving window analysis; Pinaceae; *Pinus arizonica* Engelm.; *Pinus ponderosa* var. *brachyptera* (Engelm.); Ponderosae; response function; tree rings

---

1. Introduction

The Madrean Sky Islands are rugged mountain ranges isolated by desert that signify natural ecological laboratories [1]. The steep rocky terrain offers opportunities to study biological populations across varied microclimate gradients and landscapes. Tree growth at the ecotone between communities is particularly susceptible to varying climate because small changes in the environment may have a large impact on annual growth [2,3]. In the Western United States, the consequences of rising temperatures and seasonal shifts toward earlier onset of spring (i.e., warmer temperatures) and reduced snowpack are being observed at high elevations, and these trends in shifting climate are expected to increase the length of hydrological drought by the end of the century [4].

The Sky Islands of Southeastern Arizona include the well-known Santa Catalina Mountains, which contain a habitat suitable for stands of ponderosa pine that include three morphological variants representing two distinct species [5,6]. At high elevation, two species co-occur, *Pinus arizonica* Engelm.

and the closely related *P. ponderosa* Lawson & Lawson var. *brachyptera* (Engelm.) Lemmon. The latter species is also known as Taxon X, and the taxa are clearly distinguishable by needle traits with high heritability [5–8]. *P. ponderosa* var. *brachyptera* exists in two forms, a nearly pure three-needle type that survives at the highest altitudes on southern slopes (2300–3000 m), northern slopes, and cold air drainages, and a mixed-needle type that is interspersed with the three-needle type at transition zones. *P. arizonica* is a five-needle type found at lower elevations (below 2600 m). Thus, the five-needle pine is more successful at warmer and drier, lower elevations, whereas the three-needle pine survives at colder and wetter, higher elevations. On steep southern slopes the sharp transition of species occurs over just *c.*130 m slope distance [6,9].

We combined our dataset with Hal Fritts' Mt. Bigelow chronologies [10–12], part of an earlier meta-analysis which consisted of the three co-occurring taxa (three-needle, mixed-needle, and five-needle types) sampled across a narrow climate gradient. We sampled the same population *c.* 50 years later for a comparative analysis of similar aged cohorts, and also sampled the less dry Mt. Lemmon location to determine the effect of site conditions on growth responses. Our goal was to determine whether the climate–growth relationships have changed over the last century. Shifting seasonality or relationships with limiting factors would have implications for reliably predicting the vulnerability of tree species and needle types to climate change, conservation management, and climate reconstruction.

Previously, we reported on the two species of ponderosa pine displaying different growth responses to moisture stress that varied based on the microsite environment [13]. *P. arizonica's* growth was reduced for longer periods by drought than *P. ponderosa* var. *brachyptera*, and the climatic response was greater at the site with higher soil moisture content (Mt. Lemmon). Since the turn of the twentieth century, average temperatures have been steadily increasing globally [14], and regionally (Figure 1). Because rising temperatures, variable precipitation (Figure 1) and predicted shifts in the monsoon season will influence weather patterns and drought in the Southwestern United States [15], we hypothesized that the seasonality of limiting factors will change over time. The objective of the study was to compare the correlations of ring widths with temperature and precipitation to assess shifts in the seasonality of climate–growth responses.

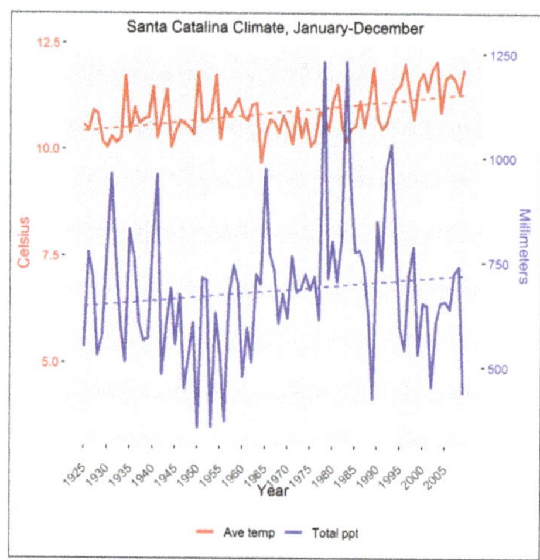

Figure 1. Climate diagram for the Santa Catalina Mountains, for the 84-year reference period spanning 1925–2009 for average annual temperature (Ave temp,°C; red line) and total annual precipitation (Total ppt, mm; blue line), respectively. The Ave temp increased by 1.3 °C over the course of the data set (from 10.6 °C to 11.9 °C). The dotted lines are the linear trend lines. Site-specific climatic data sets [Mt. Lemmon (MTL), Mt. Bigelow (BIG)] [13,16] were validated with local climate data [13], and averaged to construct this diagram.

2. Materials and Methods

2.1. Study Area

Tree-ring cores were sampled from two rugged southern slopes in the Santa Catalina Mountains located 28 km northeast of Tucson, AZ, USA: Mt. Lemmon (MTL) and Mt. Bigelow (BIG; Figure 2). BIG (32.41, −110.71, 2534 m a.s.l.) is drier than MTL (32.43, −110.79, 2577 m a.s.l.), with average available water-holding capacities of 3.8% and 9.2%, respectively [13]. The climate of the desert southwest is semiarid and warm with two rainy seasons, the summer monsoons (July through September) and winter cyclones (November through March) each delivering up to half of the annual precipitation to the region [17]. *P. arizonica*, and *P. ponderosa* var. *brachyptera* (consisting of two-needle types described below) are dominant species in the mixed conifer forest. We selected archived samples from transition zones where the three different needle types all grew together, which included the lower and upper moisture availability limits for *P. ponderosa* var. *brachyptera* and *P. arizonica*, respectively.

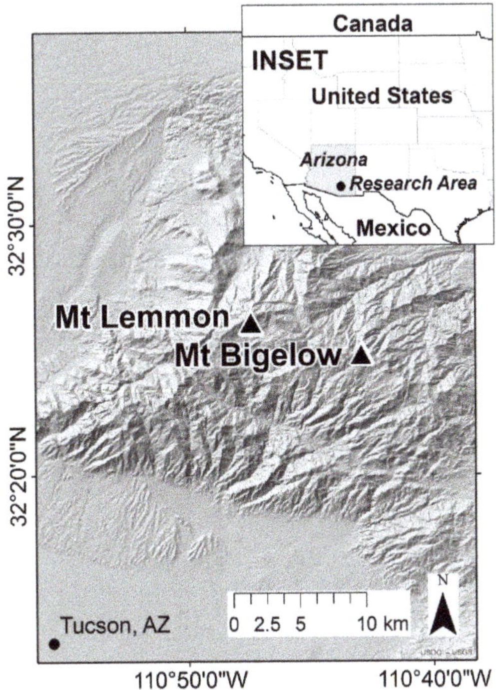

Figure 2. Locations of the two populations sampled for growth analysis: Mt. Lemmon (MTL) and Mt. Bigelow (BIG). Black symbols locate study plots northeast of Tucson, AZ where three-needle, mixed-needle, and five-needle pines are sympatric.

2.2. Climate Data

The local meteorological stations provided only short but accurate climate records; thus, specific high-resolution gridded climate (1 km) datasets were developed using the ANUSPLIN package [16,18]. Climate dataset development and validation were described earlier [13]. Climatic variables were summarized to monthly values of average temperature (TAVG), average minimum temperature (TMIN), and total precipitation (PCP) because they were previously found to be significant drivers of growth responses [13].

2.3. Tree Growth Chronologies and Data Archives

We obtained tree-ring widths for *P. arizonica* [chronology (crn) 651000] and *P. ponderosa* var. *brachyptera* (three-needle type, crn 631000; and mixed-needle type, crn 641000), located at 2500 m in elevation on south-facing slopes, from the Laboratory of Tree-Ring Research (LTRR, Tucson, AZ, USA; Table S1; Figure S1). These early period (E) data (1881–1960) from BIG were acquired as three composite chronologies that had been standardized with a negative exponential curve and converted to unit-less ring-width indices (RWI) [10–12] then trimmed to 1891–1949 for E analysis. Archived samples were not available from MTL for an early period analysis.

In addition, we developed six new composite chronologies (three from BIG, three from MTL) from raw ring widths archived with the International Tree-Ring Data Bank (ITRDB; Table S1, Figures S2 and S3) [19]. The recent period (R; 1950–2007) raw chronologies did not overlap (in time scale) with early period to simplify the response, and were de-trended using a modified negative exponential curve to remove juvenile and geometric age trends, then converted to standardized ring-width indices using the ARSTAN program [20]. The mean chronologies [computed by ARSTAN (Tucson, AZ, USA) or acquired from the LTRR] were visualized by using plotting functions in the dplR package [21,22] for R [23].

The objective of this study was to compare the recent period tree-ring data with the analysis conducted *c.* 50 years ago by Fritts [11,12], while using the same age of trees for the analysis. Expressed Population Signal (EPS) measures the common variability of a composite chronology and for early period was *c.* 0.89–0.96, indicating a strong common signal (see Appendix A for EPS thresholds). The recent chronologies comprised of a minimum of six trees of each needle type (EPS 0.82–0.90) were selected from the larger archived dataset [19]. Trees were characterized by average needle number per fascicle (Table 1) reported for the selected individuals [13]. Trees that averaged >4.6 needles per fascicle were designated *P. arizonica* [24], and *P. ponderosa* var. *brachyptera* contained two taxa identified as a mixed-needle tree (3.2 ≤ mean ≤ 4.6 needles per fascicle), and a three-needle tree (<3.2 needles per fascicle) [5,9,25].

Table 1. Average needle counts for trees sampled for tree-ring correlation analyses at two sites: Mt. Lemmon (MTL) and Mt. Bigelow (BIG). Three taxa were sampled: *Pinus ponderosa* var. *brachyptera* [three (3)-needle type], *Pinus ponderosa* var. *brachyptera* [mixed (M)-needle type], and *Pinus arizonica* [five (5)-needle type].

Site.	Taxa	Ave (#)	Range	SD	N
BIG	3	3.1	3.0–3.1	0.1	7
MTL	3	3.1	3.0–3.2	0.1	7
BIG	M	3.7	3.2–4.6	0.6	6
MTL	M	3.8	3.4–4.6	0.5	7
BIG	5	4.9	4.8–5.0	0.1	7
MTL	5	4.9	4.9–5.0	0.0	6

Ave (#): grand mean of average needle number per fascicle; Range: difference between the highest and lowest Ave (#); SD: standard deviation of Ave (#), N: number of trees sampled.

2.4. Climate–Growth Relationships

Our first analysis of climatic influence on tree-ring growth examined the correlations between climatic variables and ring-width index values using multivariate estimates obtained from the principal component regression model [12]. We used the relationship between local climatic variables and inter-annual growth to determine the differences in seasonal growth responses. We analyzed the response function correlations similarly to Marquardt et al. (2019) [13], with modifications. The analysis was conducted with standard ARSTAN chronologies, PCP, and TMIN climate variables, and needle type was the subject of analysis, with chronologies separated into two time periods: early period (E) and recent period (R). Analyses computed bootstrapped response functions using the treeclim package [26] at three site × period combinations (BIG-E, BIG-R, MTL-R). Eight seasonal climatic variables (four PCP and four TMIN) were partitioned to quantify precipitation and temperature effects during times relevant for tree growth in our study areas. Seasonal variables were derived by combining the monthly data into two rainy seasons consisting of three and five months, respectively: summer (July-September) and winter (November-March), separated by three months of arid spring (April-June). October is a transition month between the two monsoon seasons and is not included in the analysis. For specific seasons with significant response function values ($p < 0.05$), those values lying above the axis show a positive response between tree-ring width and climatic variables, and values lying below the axis show a negative response. To examine the impact (on growth) from winter precipitation, we examined one dry winter in 1961, and one normal moisture winter in 1962 across both sites, which had similar mean series lengths [13]. We used the percent difference of ring width indices to quantify the growth variability between years with contrasting moisture conditions.

To evaluate the temporal stability of climate–growth relationships, we compared climate data and composite chronologies using a univariate moving window correlation analysis within the treeclim package for R using the 'dcc' function [26]. The 18-month dendroclimatic window was set from April (previous year) to September (current year) with a 30-year moving interval and 5-year offset. Two climate variables were considered separately for analysis. Total precipitation (PCP) and average temperature (TAVG) were divided into four seasons of three months each: winter (January-March), spring (April-June), summer (July-September), and fall (October-December). The previous summer and fall seasons were also considered, which increased the number of seasons analyzed to six. Temporal instability of the moving correlation functions was tested with G-test to determine which time series fluctuations were significantly different from those of a random time series [26]. The treeclim results of the moving window analyses were plotted using functions in the 'corrplot' package [27] for R.

3. Results

3.1. Seasonal Climate–Growth Relationships

Current spring PCP was the strongest predictor of tree growth for all needle types for BIG-R (average $r = 0.27 \pm 0.04$) and MTL-R (average $r = 0.36 \pm 0.04$) after 1950. The strongest growth responses were positive, producing significant correlations with spring PCP for all needle types at both sites for the recent period (Table 2). In contrast, significant positive PCP-growth correlations were observed in summer rather than spring for BIG-E for all needle-types (average $r = 0.29 \pm 0.03$; Table 2). The response function analysis (independent of period and location) indicated that 67% (4 out of 6) of mixed- and five-needle type populations recorded a significant climate signal during winter prior to the growing season (Table 2). For the two periods, significant positive PCP-growth correlations in winter were observed only for BIG-E (average $r = 0.24 \pm 0.04$) and at MTL-R (average $r = 0.33 \pm 0.03$) and only for mixed- and five-needle types. Winter precipitation was averaged between sites for two years with contrasting moisture patterns. The winter of 1961 was one of the driest on record (137.2 mm), but the following year (1962) was of normal moisture conditions (319.43mm). Considering mean ring-width indices, growth for the five-needle type was 41% greater on average in 1962 than in 1961 (Table 3). A smaller increase in growth occurred for the three-needle type (26%) and mixed-needle type (19%).

Table 2. Seasonal precipitation-growth relationships were determined by principal components multiple regression for three taxa [three-needle (3N), mixed-needle (MN), five-needle (5N)] growing at two transition zones [Mt. Bigelow (BIG), (Mt. Lemmon (MTL)] within the Santa Catalina Mountains.

| | Early | | | | Recent | | | | Recent | | | |
| Taxa | BIG (1891–1949) 2500 m | | | | BIG (1950–2007) 2534 m | | | | MTL (1950–2007) 2577 m | | | |
	PrSum	Win	Spr	Sum	PrSum	Win	Spr	Sum	PrSum	Win	Spr	Sum
3N	-	-	-	0.32	-	-	0.32	-	-	-	0.32	-
MN	-	0.26	-	0.29	-	-	0.26	-	-	0.35	0.38	-
5N	-	0.21	-	0.26	-	-	0.24	-	-	0.31	0.39	-

PrSum = previous summer (July–September); Win = winter (November–March); Spr = spring (April–June); Sum = summer (July–September). The standard index composite chronologies analyzed were BIG-E for the early (E) period of 1891–1949, and BIG-R and MTL-R for the recent (R) period of 1950–2007. Only significant correlations are reported. All coefficients and confidence intervals are reported in Table S2.

Table 3. Average (AVE) and standard deviation (SD) of mean ring-width indices and percent difference (% DIFF) in growth between one dry winter (1961) and one normal winter (1962) for three-needle, mixed-needle, and five-needle types at two transition zones [Mt. Bigelow (BIG), (Mt. Lemmon (MTL)] within the Santa Catalina Mountains.

Three-Needle Type					
Year	BIG-R	MTL-R	AVE	SD	% DIFF
1961	731	769	750.0	026.9	
1962	1000	886	943.0	080.6	26
Mixed-Needle Type					
Year	BIG-R	MTL-R	AVE	SD	% DIFF
1961	832	673	752.5	112.4	
1962	967	828	897.5	098.3	19
Five-Needle Type					
Year	BIG-R	MTL-R	AVE	SD	% DIFF
1961	716	627	671.5	062.9	
1962	966	898	932.0	048.1	41

Growth data for each needle type were obtained from standardized chronologies developed for recent (R) period tree-ring correlation analyses.

3.2. Temporal Stability of Climate–Growth Relationships

The PCP–growth relationships were stable; the G-tests for PCP are non-significant ($p > 0.05$) for all needle types analyzed across sites and periods (BIG-E; BIG-R; MTL-R; data not shown). Because these correlation results supported the response function analysis reported above, we have only shown data for the PCP response function analysis.

The TAVG–growth relationship was stable from 1895–1949 for BIG-E across needle types (i.e., G-test p-values > 0.05; Figure 3A). The season with the largest number of individual significant negative correlations ($c. < -0.4$; $\alpha = 0.05$) was PrSum(April–June; i.e., summer of the previous year) with 4–6 correlations (out of 6), followed by Summer with 2–5 correlations (out of 6). In contrast, the relationship between TAVG and growth became unstable during the recent period from 1960 to 2004 ($p \leq 0.05$; Figure 3B,C). This instability was indicated by significant G-tests (denoted by #), and changes in sign for seasonal TAVG–growth correlations, from negative to positive for PrSum for BIG-R (Figure 3B) and from positive to negative for PrevFall (July–September; i.e., fall of the previous year) at MTL-R (Figure 3C) across all three needle types. For BIG-R, the fall season had the greatest frequency of significant positive correlations ($c. > 0.4$; $\alpha = 0.05$) with 5–6 correlations (out of 6). In comparison, the growth of MTL-R trees experienced mostly negative correlations with TAVG. The season with the largest number of significant negative correlations ($c. < -0.4$; $\alpha = 0.05$) was winter with 1–2 correlations (out of 6) for all needle types.

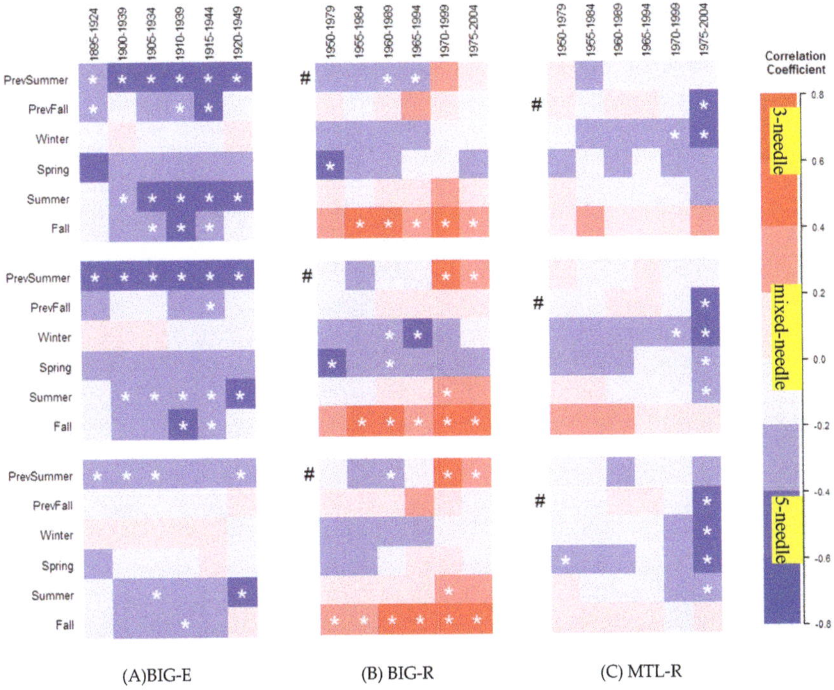

(A) BIG-E   (B) BIG-R   (C) MTL-R

Figure 3. Moving window correlation coefficients for seasonal average temperature (TAVG)–growth relationships for three different needle types, across two time periods (Early (E), Recent (R)) at two transition zones within the Santa Catalina Mountains. From top to bottom are correlations for three-needle, mixed-needle, and five-needle types, with columns (**A–C**) designating site and time period. Color scale = correlation coefficients; significant correlations ($p < 0.05$) are denoted by *. Intervals with low correlations (*c.* +0.2) are shaded pale red or pale blue, with the color intensity increasing with correlation strength. Within time periods significant G-tests are denoted by #, which indicates significant variability in the TAVG growth relationship over the time period (i.e., a sign of temporal instability). Seasons fall along the *y*-axis, which are defined as PrSum = (April–June), PrevFall = (July–September), Winter = (October–December), Spring = (January–March), Summer = (April–June), and Fall = (July–September). The standard index was used in the growth analyses. Confidence intervals are reported in Table S3.

4. Discussion

4.1. Seasonal Climate-Growth Relationships

Our chronology analysis highlighted a dramatic shift in annual growth in response to climatic changes since the late 1950s [28]. Early period Mt. Bigelow ring widths correlated most strongly with summer precipitation, indicating moisture conditions were sufficient for stomatal conductance and annual growth to occur during the entire summer season [29], thus maximizing biomass production. In comparison, recent climate–growth relationships were cool season correlations that occur when trees restrict their wood production to early spring while conditions are favorable for growth at dry sites [12,29] Thus, our recent season data indicated spring precipitation was the most important variable for tree growth, which is also a crucial time for the formation of male and female reproductive structures. These results support our hypothesis that the seasonality of climate variables important to annual growth has shifted over time. We also found that winter precipitation was correlated with the growth of mixed- and five-needle types, but not with growth of the three-needle type, suggesting the three taxa have different ecological requirements for growth at their contact zone.

We report important fundamental differences within and between species in seasonal response to climate, which suggest that adaptation of tree species to length of growing season may have occurred. These results support the influence of historical precipitation patterns and climate restrictions for different taxa [30,31]. The length of growing season of *P. ponderosa* in Nevada was controlled entirely by moisture conditions and not temperature by monitoring the cambial activity for two consecutive years with contrasting moisture, one wet winter and one dry spring [32]. Summer precipitation was not used by the trees unless the start of the growing season was delayed by dry winter and/or spring conditions. Because moisture availability determines the length of growing season for all needle types of ponderosa pine studied in the Santa Catalina Mountains [13], we hypothesize that one taxon may optimize growth over a shorter or longer growing season (five-needle, three-needle, respectively) [10], and a third taxon may grow best under a variable growing season (mixed-needle), which we call the dry cool-season hypothesis.

In support of this dry cool-season hypothesis, cambial activity was measured during the dry 1961 growing season [10]; the average Palmer Drought Severity Index (PDSI), a measure of soil moisture availability [33], was c. −3.0 [13]. [More negative values of PDSI indicate drought stress and range from -4 to +4]. The cambial expansion of *P. ponderosa* var. *brachyptera* (the three- and mixed-needle types) initiated under dry winter conditions in February, followed by the five-needle type (*P. arizonica*) six weeks later in April. Cambial growth occurred for all needle types during the summer, and cessation of growth coincided in September. All needle types co-existed at the Mt. Bigelow transition zone (2500 m a.s.l.) including trees that were analyzed for climatic response in this study and the prior study [11]. Although the initiation of radial growth for the following normal moisture year (average PDSI *c.* −0.5) [13] was not determined, we can predict the growth response of the five-needle type to wet winter conditions from the recent literature on moisture driven cambial activity described above [32]. Because the five-needle type responds positively to winter precipitation, the taxon should initiate growth earlier. Further support was provided by our chronology data, which showed that growth for the five-needle type is robust when winter moisture conditions are normal (compared to dry winter). Also, a 15–22% increase in growth for the five-needle type was observed above that for the three- and mixed-needle types, respectively, for normal moisture winter. This increase in growth is suggestive that five-needle type starts cambial expansion earlier in wet springs than dry springs and supports our earlier study [13] that found the five-needle pine responded positively to PDSI for longer periods of time. Long correlation with PDSI suggests a possible adaptation to drought under a mechanism that 5-needle trees grow only when there is a reliable water source. Therefore, our study results reinforce that moisture availability determines the length of growing season of ponderosa pine, and that some needle types may display variability in the timing of xylogenesis [32].

4.2. Temporal Stability of Climate–Growth Relationships

We found opposing growth trends in response to temperature after 1950 between the two sites, which indicates that combining tree-ring data from multiple sites into a single chronology can dilute the individual climate signals. Opposing temperature–growth trends were first reported in white spruce stands occurring at the tree-line in two mountain ranges sampled across Alaska, and that temperature explained more variability in annual tree growth post-1950 [34]. These regional Alaskan forests are composed of individual trees growing in heterogeneous environments where temperature thresholds operate, influencing tree biology and the response to limiting factors that appears in the final chronology. Our report extends these tree-line studies to semiarid regions where precipitation rather than temperature is the factor most limiting to tree annual growth. Nonetheless, we also observed a significant shift from positive to negative growth correlations with temperature under less dry conditions for five-needle and mixed-needle types during cool seasons, which is indicative of temperature-induced drought stress [34], or growing season respiration [13] limiting the ponderosa pine annual growth. In comparison, at the drier site we observed a shift from negative to positive temperature correlations during PrSum season post-1950, indicating reduced water stress. Furthermore, if unaccounted for, these opposing trends to temperature are weakening the warming climate signal (post-1950) by combining positive and negative growth responses that adversely affect growth response analyses and by extension climate reconstructions based on ring-width data, even in dry environments where precipitation is growth limiting.

Also, by extending our analysis to include the fall season, we observed that growth in individual trees is positively and significantly correlated to temperature. These results suggest that all needle types may respond positively to a range of temperature increases late in the season at the contact zone of dry habitats. However, an extended growing season most likely will be determined by moisture availability and not rising temperatures, as recently described for *P. ponderosa*, a closely related species [32].

5. Conclusions

A temporal shift in limiting factor indicated spring is the most important growing season for recent period; thus, the three taxa that responded to summer precipitation pre-1950 are now responding to spring precipitation post-1950. This shift in limiting factor represents a shift in allocation of tree resources from maximizing biomass (summer) to reproduction (spring), a critical time for tree phenology and the formation of male and female reproductive structures. Five- and mixed-needle types correlated positively to winter precipitation, suggesting soil moisture was controlling the length of growing season. Also, we found significant fluctuations in temperature–growth correlations during recent period for all needle types. Thus, a shift in limiting factor that impacts the growth sensitivity of trees, such as warming trends that increase local moisture stress, may also impact the stability of climate–growth relationships, and if unaccounted for, obscures the use of tree-rings to analyze temperature–growth responses and reconstruct climate on moisture-limited sites.

Supplementary Materials: The following are available online at http://www.mdpi.com/1999-4907/10/11/1011/s1, Table S1: Ponderosa pine tree-ring data archives, Table S2: Coefficients and confidence intervals for Table 2's response function seasonal precipitation–growth analysis, Table S3: Coefficients and confidence intervals for Figure 3's moving window seasonal TAVG growth analysis, Figure S1: Average yearly standard chronology for three-needle, mixed-needle, and five-needle types in the early BIG tree-ring study [10,11], Figure S2: Average yearly standard chronologies for three-needle, mixed-needle, and five-needle types in the recent BIG tree-ring study, Figure S3: Average yearly standard chronologies for three-needle, mixed-needle, and five-needle types in the recent MTL tree-ring study.

Author Contributions: P.E.M. conceived the study, designed the experiments, and analyzed data; B.R.M. assisted with data analysis; F.W.T. contacted the LTRR to obtain historical tree-ring data files; P.E.M. wrote the paper; F.W.T. and B.R.M. edited the paper.

Funding: The USDA Forest Service, Northern Research Station, and the Department of Plant Biology, Michigan State University provided financial support for this research project.

Acknowledgments: This paper is part of a dissertation submitted to Michigan State University in partial fulfillment of requirements for a Doctor of Philosophy degree. We thank the following people for help on the project: M. Munro and C. Baisan supplied H. Fritts' chronology samples. J. Stanovick provided statistical support. H. Jenson provided GIS support. D. Donner reviewed the paper.

Conflicts of Interest: The authors declare no conflict of interest.

Appendix A

Defining Chronology Confidence and EPS Thresholds

Analysis of variance (ANOVA) and correlation-derived estimates of population parameters require a minimum of five single-core series and 30-annual rings to provide accurate values for tree-ring statistical analysis [35]. Therefore, we selected 6–7 trees (two or more cores per tree), setting a minimum length limit of a 47-year common interval to construct composite chronologies for each group of trees (Table S1; Figures S2 and S3). The percent variance component for the group chronology VC(Y) as determined by analysis of variance is equivalent to the between tree correlation $\bar{r}_{bt}$ of correlation tree-ring analysis [35,36]. $\overline{R}_{bt}$ measures the common signal strength and is dependent on sample depth [37]. Expressed Population Signal (EPS) is a measure of statistical quality to help produce a reliable population level estimate of the climate signal, and is dependent on $\overline{r}_{bt}$ [35]

$$EPS = \frac{t\,\overline{r}_{bt}}{t\,x\,\overline{r}_{bt} + (1 - \overline{r}_{bt})}$$

Although the minimum EPS value of 0.85 is recommended by Wigley et al. [36] to be routinely applied in dendrochronological analyses, we considered EPS values ranging from 0.82 to 0.90 for each group of trees that included growth data from 1950 to 2007, which is a good reflection of overall chronology confidence. Given that our study design includes a variable core depth, this method of estimating the common chronology signal with a range of EPS values was simpler to apply. The chronology signal describes the variance in common to all tree-ring series sampled and analyzed at a specific site. This approach of using a range of EPS values was supported by Briffa and Jones' [34] suggestion that a minimum value may not represent the most appropriate estimate of chronology confidence for all situations. For example, a 3% reduction in EPS from 0.85 to 0.82 would signify a reduction of 2% in explained climate variance. Thus, for climate–growth relationships with 64–65% of variance attributed to climate, as reported by Fritts [11] for the mixed- and three-needle types of *Ponderosae* on Mt. Bigelow, an EPS threshold of 0.85 would have reduced the explained climate variance to 55% (64.5% × 0.85). In comparison, an EPS threshold of 0.82 would have reduced the explained variance to 53%. Therefore, the additional chronology error would have restricted the explained climate variance by just 2%. Thus, it seems unlikely that an additional 3% reduction in chronology confidence (from 0.85 to 0.82) would affect the estimation of the climate signal within needle types of standardized chronologies, when sensitivities to climate are expected to be high, as in this study [10,11,13,36].

The EPS thresholds for Fritts' [11] study can be estimated from formulas described by Wigley et al. [37] as:

$$SNR = \frac{[N(\%Y)]}{100 - \%Y}$$

$$EPS = \frac{SNR}{1 + SNR}$$

where SNR is the signal to noise ratio, N is the number of series, and %Y is the percent common variance [35]. As an example, when substituting 12 for N and 41 for %Y [11] (%EMS, from Table 1), we derived SNR = 8.4 and EPS = 0.89 for the five-needle type of the prior Mt. Bigelow study. Similarly, for the three-needle type, by substituting 65 for %Y [11] we derive an EPS value of 0.96. These values approximate EPS of the current study, as 11 years of data were removed (1950 to 1960) from the chronologies used for the early period analysis (BIG-E).

References

1. Warshall, P. *The Madrean sky island archipelago: A planetary overview. In Biodiversity and Management of the Madrean Archipelago: The Sky Islands of Southwestern United States and Northwestern Mexico*; DeBano, L.F., Ffolliott, P.F., Ortega-Rubio, A., Gottfried, G.J., Hamre, R.H., Edminster, C.B., Eds.; US Department of Agriculture: Fort Collins, CO, USA, 1995; pp. 6–18. [CrossRef]
2. Jump, A.S.; Hunt, J.M.; Penuelas, J. Rapid climate change-related growth decline at the southern range edge of Fagus sylvatica. *Glob. Chang. Biol.* **2006**, *12*, 2163–2174. [CrossRef]
3. Tegel, W.; Seim, A.; Hakelberg, D.; Hoffmann, S.; Panev, M.; Westphal, T.; Büntgen, U. A recent growth increase of European beech (*Fagus sylvatica* L.) at its Mediterranean distribution limit contradicts drought stress. *Eur. J. Forest Res.* **2014**, *133*, 61–71. [CrossRef]
4. USGCRP. *Climate Science Special Report: Fourth National Climate Assessment*; Wuebbles, D.J., Fahey, D.W., Hibbard, K.A., Dokken, D.J., Stewart, B.C., Maycock, T.K., Eds.; U.S. Global Change Research Program: Washington, DC, USA, 2017; Volume 1, p. 470. [CrossRef]

5. Rehfeldt, G.E. Systematics and genetic structure of *Ponderosae* taxa (Pinaceae) inhabiting the Mountain Islands of the Southwest. *Am. J. Bot.* **1999**, *86*, 741–752. [CrossRef] [PubMed]

6. Epperson, B.K.; Telewski, F.W.; Willyard, A. Chloroplast diversity in a putative hybrid swarm of *Ponderosae* (Pinaceae). *Am. J. Bot.* **2009**, *96*, 707–712. [CrossRef] [PubMed]

7. Willyard, A.; Gernandt, D.S.; Potter, K.; Hipkins, V.; Marquardt, P.; Mahalovich, M.F.; Langer, S.K.; Telewski, F.W.; Cooper, B.; Douglas, C.; et al. *Pinus ponderosa*: A checkered past obscured four species. *Am. J. Bot.* **2017**, *104*, 161–181. [CrossRef] [PubMed]

8. Rehfeldt, G.E. Early selection in *Pinus ponderosa*: Compromises between growth potential and growth rhythm in developing breeding strategies. *Forest Sci.* **1992**, *38*, 661–677. [CrossRef]

9. Epperson, B.K.; Telewski, F.W.; Plovanich-Jones, A.E.; Grimes, J.E. Clinal differentiation and putative hybridization in a contact zone of *Pinus ponderosa* and *P. arizonica* (Pinaceae). *Am. J. Bot.* **2001**, *88*, 1052–1057. [CrossRef] [PubMed]

10. Dodge, R.A. Investigations into the Ecological Relationships of Ponderosa Pine in Southeast Arizona. Ph.D. Thesis, University of Arizona, Tucson, AZ, USA, 1963.

11. Fritts, H.C. Computer programs for tree-ring research. *Tree-Ring Bull.* **1963**, *25*, 2–7.

12. Fritts, H.C. Relationships of ring widths in arid-site conifers to variations in monthly temperature and precipitation. *Ecol. Monogr.* **1974**, *44*, 411–440. [CrossRef]

13. Marquardt, P.E.; Miranda, B.R.; Jennings, S.; Thurston, G.; Telewski, F.W. Variable climate response differentiates growth of the Sky Island ponderosa pines. *Trees* **2019**, *3*, 317–332. [CrossRef]

14. IPCC. 2007: Climate Change. The physical science basis. In *Contribution of Working Group I to the Fourth Assessment Report of the Intergovernmental Panel on Climate Change*; Solomon, S., Qin, D., Manning, M., Chen, Z., Marquis, M., Averyt, K.B., Tignor, M., Miller, H.L., Eds.; Cambridge University Press: Cambridge, UK; New York, NY, USA, 2007.

15. Pascale, S.; Boos, W.R.; Bordoni, S.; Delworth, T.L.; Kapnick, S.B.; Murakami, H.; Vecchi, G.A.; Zhang, W. Weakening of the North American monsoon with global warming. *Nat. Clim. Chang.* **2017**, *7*, 806–812. [CrossRef]

16. McKenney, D.W.; Hutchinson, M.F.; Papadopol, P.; Lawrence, K.; Pedlar, J.; Campbell, K.; Milewska, E.; Hopkinson, R.F.; Price, D.; Owen, T. Customized spatial climate models for North America. *Bull. Am. Meteorol. Soc.* **2011**, *92*, 1611–1622. [CrossRef]

17. Sheppard, P.R.; Comrie, A.C.; Packin, G.D.; Angersbach, K.; Hughes, M.K. The climate of the US Southwest. *Clim. Res.* **2002**, *21*, 219–238. [CrossRef]

18. Hutchinson, M.; Xu, T. *Anusplin Version 4.4 User Guide*; Australia National University: Canberra, Australia, 2013.

19. Marquardt, P.E.; Miranda, B.R.; Telewski, F.W. (2018-01-18): NOAA/WDS Paleoclimatology - Mt. Bigelow – PIAZ, PIPO – ITRDB AZ595-596 [Study 23650-23651]; Mt. Lemmon – PIAZ, PIPO – ITRDB AZ597-598 [Study 23652-23653]. NOAA National Centers for Environmental Information. Available online: https://www.ncdc.noaa.gov/paleo/study/xxxxx (accessed on 1 October 2018).

20. Cook, E.R. ATime Series Analysis Approach to Tree Ring Standardization. Ph.D. Thesis, University of Arizona, Tucson, AZ, USA, 1985.

21. Bunn, A.G. A dendrochronology program library in R (dplR). *Dendrochronologia* **2008**, *26*, 115–124. [CrossRef]

22. Bunn, A.G. Statistical and visual crossdating in R using the dplR library. *Dendrochronologia* **2010**, *28*, 251–258. [CrossRef]

23. R Core Team. *R: A Language and Environment for Statistical Computing*; R Foundation for Statistical Computing: Vienna, Austria, 2016; Available online: https://www.R-project.org/ (accessed on 1 October 2018).

24. Peloquin, R.L. The identification of three-species hybrids in the ponderosa pine complex. *Southwest. Nat.* **1984**, *29*, 115–122. [CrossRef]

25. Rehfeldt, G.E. Genetic variation in the *Ponderosae* of the southwest. *Am. J. Bot.* **1993**, *80*, 330–343. [CrossRef]

26. Zang, C.; Biondi, F. Treeclim: An R package for the numerical calibration of proxy-climate relationships. *Ecography* **2015**, *38*, 431–436. [CrossRef]

27. Wei, T.; Simko, V. R Package "Corrplot": Visualization of a Correlation Matrix (Version 0.84). 2017. Available online: https://github.com/taiyun/corrplot (accessed on 1 October 2018).

28. Kienast, F.; Luxmoore, R.J. Tree-ring analysis and conifer growth responses to increased atmospheric CO_2 levels. *Oecologia* **1988**, *76*, 487–495. [CrossRef] [PubMed]

29. Lévesque, M.; Rigling, A.; Bugmann, H.; Weber, P.; Brang, P. Growth response of five co-occurring conifers to drought across a wide climatic gradient in Central Europe. *Agri. Forest Meteorol.* **2014**, *197*, 1–12. [CrossRef]

30. Kilgore, J. Distribution and Ecophysiology of the *Ponderosae* in the Santa Catalina Mountains of Southern Arizona. Ph.D. Thesis, Michigan State University, East Lansing, MI, USA, 2007.

31. Norris, J.R.; Jackson, S.T.; Betancourt, J.L. Classification tree and minimum-volume ellipsoid analyses of the distribution of ponderosa pine in the western USA. *J. Biogeogr.* **2006**, *33*, 342–360. [CrossRef]

32. Ziaco, E.; Truettner, C.; Biondi, F.; Bullock, S. Moisture-driven xylogenesis in *Pinus ponderosa* from a Mojave Desert mountain reveals high phenological plasticity. *Plant Cell Environ.* **2018**, *41*, 823–836. [CrossRef] [PubMed]

33. Palmer, W.C. *Meteorological Drought*; U.S. Weather Bureau Research Paper No. 45; US Department of Commerce: Washington, DC, USA, 1965; Volume 30, pp. 1–58.

34. Wilmking, M.; Juday, G.P.; Barber, V.A.; Zald, H.S. Recent climate warming forces contrasting growth responses of white spruce at treeline in Alaska through temperature thresholds. *Glob. Chang. Biol.* **2004**, *10*, 1724–1736. [CrossRef]

35. Briffa, K.; Jones, P. Basic Chronology statistics and assessment. In *Methods of Dendrochronology*; Cook, E.R., Kairiukstis, L.A., Eds.; Springer Science Business Media B.V: Dordrecht, The Netherlands, 1990; pp. 137–152. ISBN 978-90-481-4060-2.

36. Fritts, H.C. *Tree Rings and Climate*; Academic Press Inc.: London, UK, 1976; ISBN 0 12 268450-8.

37. Wigley, T.M.L.; Briffa, K.R.; Jones, P.D. On the average value of correlated time series, with applications in dendroclimatology and hydrometeorology. *J. Clim. Appl. Meteorol.* **1984**, *23*, 201–213. [CrossRef]

Article

Thermal Time and Cardinal Temperatures for Germination of *Cedrela odorata* L.

Salvador Sampayo-Maldonado [1], Cesar A. Ordoñez-Salanueva [1], Efisio Mattana [2], Tiziana Ulian [2], Michael Way [2], Elena Castillo-Lorenzo [2], Patricia D. Dávila-Aranda [3], Rafael Lira-Saade [3], Oswaldo Téllez-Valdéz [3], Norma I. Rodriguez-Arevalo [3] and Cesar M. Flores-Ortíz [1,4,*]

[1] Plant Physiology Laboratory, UBIPRO, FES Iztacala, UNAM, Tlalnepantla 54090, Estado de Mexico, Mexico; ssampayom@hotmail.com (S.S.-M.); caos@unam.mx (C.A.O.-S.)

[2] Royal Botanic Gardens, Kew, Wellcome Trust Millennium Building, Ardingly, West Sussex RH17 6TN, UK; E.Mattana@kew.org (E.M.); t.ulian@kew.org (T.U.); m.way@kew.org (M.W.); E.CastilloLorenzo@kew.org (E.C.-L.)

[3] Natural Resources, UBIPRO, FES Iztacala, UNAM, Tlalnepantla 54090, Estado de Mexico, Mexico; pdavilaa@unam.mx (P.D.D.-A.); rlira@unam.mx (R.L.-S.); tellez@unam.mx (O.T.-V.); isela.unam@gmail.com (N.I.R.-A.)

[4] National Laboratory in Health, FES Iztacala, UNAM, Tlalnepantla 54090, Estado de Mexico, Mexico

* Correspondence: cmflores@unam.mx; Tel.: +52-555-623-1137 or +52-552-922-2373

Received: 21 August 2019; Accepted: 24 September 2019; Published: 26 September 2019

Abstract: Thermal time models are useful to determine the thermal and temporal requirements for seed germination. This information may be used as a criterion for species distribution in projected scenarios of climate change, especially in threatened species like red cedar. The objectives of this work were to determine the cardinal temperatures and thermal time for seeds of *Cedrela odorata* and to predict the effect of increasing temperature in two scenarios of climate change. Seeds were placed in germination chambers at constant temperatures ranging from 5 ± 2 to 45 ± 2 °C. Germination rate was analyzed in order to calculate cardinal temperatures and thermal time. The time required for germination of 50% of population was estimated for the current climate, as well as under the A2 and B2 scenarios for the year 2050. The results showed that base, optimal and maximal temperatures were −0.5 ± 0.09, 38 ± 1.6 and 53.3 ± 2.1 °C, respectively. Thermal time ($\theta_1(50)$) was 132.74 ± 2.60 °Cd, which in the current climate scenario accumulates after 5.5 days. Under the A2 scenario using the English model, this time is shortened to 4.5 days, while under scenario B2, the time is only 10 hours shorter than the current scenario. Under the German model, the accumulation of thermal time occurs 10 and 6.5 hours sooner than in the current climate under the A2 and B2 models, respectively. The seeds showed a wide range of temperatures for germination, and according to the climate change scenarios, the thermal time accumulates over a shorter period, accelerating the germination of seeds in the understory. This is the first report of a threshold model for *C. odorata*, one of the most important forest species in tropical environments.

Keywords: *Cedrela odorata*; seeds; germination; cardinal temperatures; thermal time; climate change

---

1. Introduction

Cedrela odorata L. is a tree of the Meliaceae family and is native to the American tropics [1]. According to Mendizábal-Hernández et al. [2], it is distributed in warm to subwarm climates in Mexico and requires fertile soil with good drainage; as such it is mainly found in the humid and subhumid tropics, associated with tropical deciduous forest, tropical semi-deciduous forest, and tropical rainforest, in addition to humid montane forest. According to Romo-Lozano et al. [3], *C. odorata* provides ecosystem services to mitigate the impacts of climate change, serving as a carbon sink that maintains the resilience of ecosystems and as a source of genetic resources for medicinal and other uses in future environments. Navarro et al. [4] and Ramírez-García et al. [5] recognize its ecological importance as a pioneer species in tropical forests. It is important for reforestation of degraded areas and has potential for establishing plantations. According to Pérez-Salicrup and Esquivel [6] and Sampayo-Maldonado et al. [7], it is the most economically viable species for establishing commercial plantations, the most important for the forestry industry in Mexico, and preferred on the international market. It has a wide range of uses in construction, carpentry and cabinetmaking and therefore represents one of the most planted species in the country [8].

C. odorata is included in the Convention on International Trade in Endangered Species [9]. It is also included on the red list of threatened species in the category of vulnerable and is subject to special protection under the Mexican law Official Mexican Standard 059 (NOM-059-SEMARNAT) [10]. The distribution of plant communities is determined by several factors, but the main factor is climate. According to Gómez-Díaz et al. [11], the different models of climate change for Mexico, predict an increase in temperature and decrease in precipitation for the years 2050 and 2100. Hartmann et al. [12] mention that temperature is the most important factor in the adaptation of a species, considering its geographic origin. González [13] and Dewan et al. [14], state that temperature influences the physiological processes of seed germination including speed and germination percentage. Reduced germination time means higher economic efficiency in the production of a larger number of plants [15].

The germination response to temperature can be characterized through the germination rate and is defined by three cardinal or threshold temperatures: a base temperature (Tb) below which germination does not proceed; an optimal temperature (To) at which the rate of germination is highest; and a maximum or ceiling temperature (Tc) above which germination ceases [16]. The cardinal temperatures are the limits, while the optimum temperature is the one by which germination is fastest [17]. Thermal time is the thermal sum accumulated per day that is necessary for the germination of 50% of the seed lot [18]. Temperature is one of the most important bioclimatic elements in determining the response of seeds to changing environmental conditions [19].

According to Calzada-López et al. [20], characterizing cardinal temperatures and thermal time is useful for finding the optimal temperature for fastest germination, as a criterion for species distribution under different climate change scenarios. Ruíz-Corral et al. [21] mention that cardinal temperatures are variable among species, among populations, and even within a species, as a direct effect of adaptation. Studies of the effect of temperature on germination are therefore needed, since this is a requisite step for forest conservation and sustainable management [22]. It is important to mention that there are no previous studies of cardinal temperatures and thermal time in *C. odorata*.

Seed germination is the most vulnerable and crucial stage in the life history of a tree [23]. According to Sánchez-Monsalvo et al. [24] and Castellanos-Acuña et al. [25], germplasm that is of known genetic origin and is adapted to environmental conditions in each region will be needed in studies to understand the impact of cardinal temperatures and thermal time during germination. Such studies are required to identify the effect of increased temperature on the species potential distribution under different climate change scenarios. However, there is currently no information on the optimal temperature or thermal time for germination in tropical trees of interest to forestry in Mexico. Thus, the objectives of this research were to determine the cardinal temperatures and thermal time of *C. odorata* seeds and to predict the effect of increasing temperature under two scenarios of climate change.

2. Materials and Methods

2.1. Seed Collection

The seeds of *C. odorata* are flat ovoid, with a brown-colored testa, provided with a dark wing contained in a woody dehiscent capsule. Each capsule has between 25 to 35 seeds. Each seed has two large and flat cotyledons; the endosperm is thin and attached to the white embryo. The embryo is axial, straight and spatulate, white to cream colour, it has a short radicle which protrudes laterally.

C. odorata seeds have a wide variation in morphometric characteristics, associated with environmental and genetic factors. One kg contains approximately 94,965 ± 8108 seeds [26].

Mature *C. odorata* capsules were collected in April 2018 (2 kg, 149,406 seeds) from 15 trees as a representative sample from a population in Zozocolco de Hidalgo. The brownest capsules were collected just before dehiscence by using pruners over a tarpaulin when capsules were low enough in the tree, but frequently it was necessary to climb the tree to obtain the most mature capsules. These were put in cotton bags and kept in the laboratory at ambient temperature. All capsules opened naturally in less than one week. The seeds were manually separated from plant debris and stored in paper bags at 15 °C until germination tests were carried out.

The municipality of Zozocolco de Hidalgo is in the state of Veracruz (649265.14 E, 2223834.28 N; 183 m a.s.l.), in the Totonacapan region. The climate at the site is warm and subhumid, with rains year-round ((Af) according to García [27]), with a mean annual precipitation of 2233 ± 33.93 mm and mean temperature of 23.4 ± 3.32 °C (Figure 1). The soils are Acrisol, deep with good drainage, with a sandy clay texture and pH of 5.7 [28].

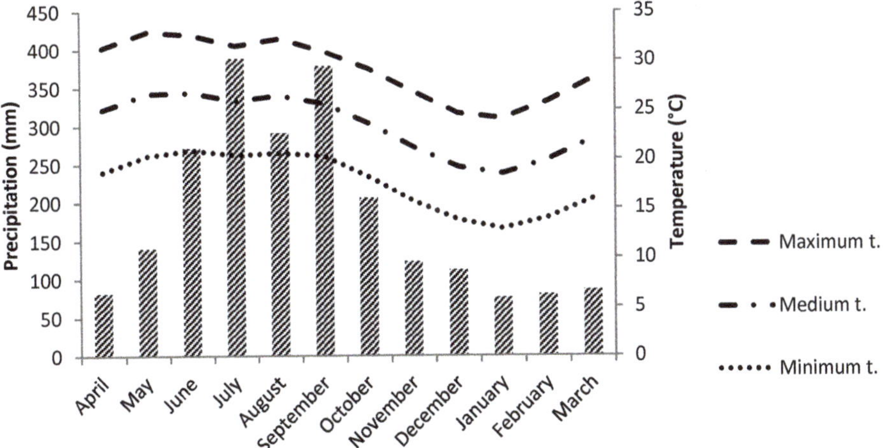

Figure 1. Climate data from the area where seeds were collected. Historical monthly average of 1981–2010. Available in global climate data (http://es.climate-data.org/).

2.2. Disinfection

Following Sampayo-Maldonado et al. [29], the seeds were treated with powder soap and running water for 10 minutes, then rinsed three times with distilled water. They were soaked in Captan® (1 g L^{-1}) for 60 minutes, then rinsed three times with distilled water. Later, they were treated with 70% ethanol for 1 minute, rinsed three times with distilled water, and finally, soaked for one minute in 6% sodium hypochlorite and rinsed three times with distilled water.

2.3. Initial Germination Test

In order to know the quality of the seed lot, an initial germination test was performed. The seeds were placed randomly on agar medium (10 g L^{-1}) in Petri dishes (6 cm diameter), in a completely randomized design with four replicates, using 25 seeds each. The Petri dishes were placed in a climate-controlled chamber at a temperature of 25 ± 2 °C, and a photoperiod of 12 light hours and 12 dark hours. The test began on May 14, 2018. A seed was considered to have germinated when the size of the radicle reached ≥2 mm [30]. A cut test was undertaken at the end of the germination test on those seeds that did not germinate to check viability. Seeds with a filled, firm and white embryo were considered viable.

The seeds were tested for germination in the Plant Physiology Laboratory at the Biotechnology and Prototype Research Unit of the Faculty of Higher Education, Iztacala of the National Autonomous University of Mexico (Laboratorio de Fisiología vegetal, Unidad de Investigación de Biotecnología y Prototipos, Facultad de Estudios Superiores Iztacala, Universidad Nacional Autónoma de México) in Tlalnepantla, state of Mexico.

2.4. Effect of Temperature on Germination

Seeds were sown under sterile conditions in a laminar flow cabinet (Novatech, Mod. CF-13). A total of 30 seeds were placed in each Petri dish, with five replicates. The dishes were then sealed with parafilm and labeled. They were then placed in germination chambers at the following constant temperatures: $5 \pm 2, 10 \pm 2, 15 \pm 2, 20 \pm 2, 25 \pm 2, 30 \pm 2, 35 \pm 2, 40 \pm 2$ and $45 \pm 2\,°C$, under a photoperiod of 12 hours light, 12 hours darkness, using halogen lamps at a light intensity of 28.05 $\mu mol\ m^{-2}\ s^{-1}$ (Quantum Meter Apogee Mod. QMSW-SS). Seeds were sown on June 1, 2018; the seeds had three weeks of storage. Counts were conducted daily for the following 61 days in order to determine the proportion of germinated seeds. Following ISTA [30] and Parmoon et al. [31], a seed was considered to have germinated when its radicle reached ≥ 2 mm, measured with a stainless-steel Vernier ruler (Truper).

2.5. Variables Evaluated

2.5.1. Total Germination

The number of germinated seeds in each Petri dish of the five repetitions of each treatment was recorded, to obtain averages in each treatment. The proportion of germinated seeds was calculated using the following equation [29]:

$$G(\%) = \frac{n}{N} * 100 \tag{1}$$

where n is the number of seeds germinated and N the total number of seeds.

2.5.2. Median Germination Time (t_{50})

The total number of days between imbibition and 50% of the total germination was recorded. Following Ordoñez-Salanueva et al. [32] a sigmoid curve was fitted to the accumulated germination, allowing the median germination time to be determined by interpolation.

2.5.3. Germination Rate

This is an estimate of the number of germinated seeds per day, according to the following equation [20]:

$$VG = \frac{G_1}{N_1} + \frac{G_2}{N_2} + \cdots + \frac{G_i}{N_i} + \cdots + \frac{G_n}{N_n} = \sum_{i=1}^{n} \frac{G_i}{N_i} \tag{2}$$

where G_i is the number of germinated seeds and N_i is the number of days after the beginning of the experiment.

2.5.4. Base Temperature (Tb)

This represents the temperature in degrees Celsius below which germination does not occur; it was calculated in percentiles, in 10% intervals, for all temperature treatments. The inverse time to germination was then graphed as a function of temperature in order to observe data trends and locate the inflection point and determine sub-optimal temperatures. Following Ellis et al. [17] a linear regression was performed to obtain the parameters for each germination percentage. The mean value of x-intercept ($\beta 0$) was then calculated and used to generate a second linear regression for each germination percentage. The mean $\beta 0$ was the base temperature.

2.5.5. Upper Threshold Temperature (Tc)

Following Hardegree [33], supra-optimal temperatures in degrees Celsius were determined for the upper threshold temperature, which were used to generate a linear regression to obtain the parameters for each germination percentage. The mean value of the x-intercept ($\beta0$) was obtained and used to do a second linear regression. The mean $\beta0$, again calculated from the second regression, represents the upper threshold temperature.

2.5.6. Optimal Temperature (To)

This represents the optimum germination temperature in degrees Celsius and was obtained by equating the straight-line equations of Tb and Tc at their intersection [33].

2.5.7. Thermal Time (θ, °C)

The inverse of the slope of the regression lines for each fraction were calculated separately to estimate the thermal time (θ, °Cd) in the sub-optimal (θ_1) and supra-optimal (θ_2) temperature range. The thermal time means the thermal requirements in temperature by time to reach the germination of each percentile in the population. Percentage data were transformed using Probit Analysis in Genstat (version 11.1.0.1504, International Ltd, Hemel Hempstead, Herts, England, UK). Linear regression for the sub- and supra-optimal temperature range for each species was used to express probit (G) as a function of θ.

To estimate the probability that a seed will germinate in a given time, the number of germinated seeds for the percentiles was calculated, at intervals of 10% for each of the temperature treatments. Then the inverse of the slope of the regression lines for each percentile was calculated and the percentage data were transformed into probits. For the sub-optimal temperature range, R^2 values were highest and residual variances were smallest when probit values were expressed as a function of θ_1 (sub-optimal thermal time). For the sub-optimal temperature range the following equation describes the form of cumulative germination response of seeds [34]:

$$Probit\ (G) = K + [\theta_1/\sigma] \tag{3}$$

where K is an intercept constant when thermal time is zero and σ is the standard deviation of the response to thermal time θ_1. The same equation was used to determine the thermal time of germination of 50% ($\theta_1(50)$) of the population.

For the supra-optimal temperature range, R^2 values were highest and residual variances were smallest when probit values were expressed as a function of θ_2 the following equation describes the form of accumulative germination response of seeds [34]:

$$Probit\ (G) = Ks + (T + \theta_2/t(G))/\sigma \tag{4}$$

where Ks is an intercept constant, when thermal time is zero, $(T + \theta_2/t(G))$ is the maximum temperature (Tc), and σ is the standard deviation of the maximum temperature (Tc). The same equation was used to determine the thermal time required for germination of 50% ($\theta_2(50)$) of the population.

2.5.8. Climate Change Scenarios

Following Gutiérrez and Trejo [35], we used the projections of mean temperature proposed by two general circulation models—the German model (MPIECHAM5) and the English model (UKMOHADGEM1)—available from the Digital Climate Atlas of Mexico [36]. The projection for the year 2050 under the A2 (severe, or pessimistic scenario with high emissions of greenhouse gases) and the B2 (conservative or non-pessimistic model with low greenhouse gas emissions) were used. The area was located on the map and the mean temperature data for the A2 and B2 scenarios was obtained. The mean temperatures were projected for the month of April, since this is the period when *C. odorata* seed dispersal occurs [1].

Each scenario was used to predict the time in which seeds of *C. odorata* accumulate the thermal time necessary for germination of 50% of the seed bank in the understory. Following Flores-Magdaleno et al. [37], the analysis was performed using the mean temperature, so that the following formula was used [38]:

$$Thermal\ sum\ (°Cd) = (Env\ T_m - T_b)t_m \tag{5}$$

where °Cd is the degrees days accumulated, T_b is the minimum germination temperature, T_m is the monthly mean temperature, and t_m is the number of days in the month.

2.6. Experimental Design and Analyses

We used a completely randomized design. All treatments were performed in similar conditions as can be seen in Figure 2. The data did not fulfill the assumptions of normality, so prior to the analysis of variance (ANOVA), percentage variables (Y) were transformed using the arcsine square root function (with the original value expressed as a proportion) [T = arcsine($\sqrt{Y}$)] [39–41]. The ANOVAs were carried out in SAS statistical software (Cary, NC, USA) [42], and the Tukey test was performed for multiple comparisons to determine significant ($p \leq 0.05$) differences between treatments. ($p \leq 0.05$).

Figure 2. The experimental unit: (**a**) Petri dish with 30 seeds and (**b**) 5 replicates.

3. Results

The seeds did not show dormancy, so no pre-germination treatments were carried out. In the initial germination test, the seed lot used had a 98% germination rate at 25 ± 2 °C.

3.1. Germination

There was a significant effect of temperature on final percentage of germination ($F_{8,36}$ 49.82 $p < 0.02$). Seeds of *C. odorata* began to germinate after three days, and at the lowest temperature they began at 28 days after sowing. Mean germination was above 50%, and the highest frequency was 90% germination (Figure 3). The highest germination percentage was at 20 ± 2 °C and the lowest percentage was at the highest temperature.

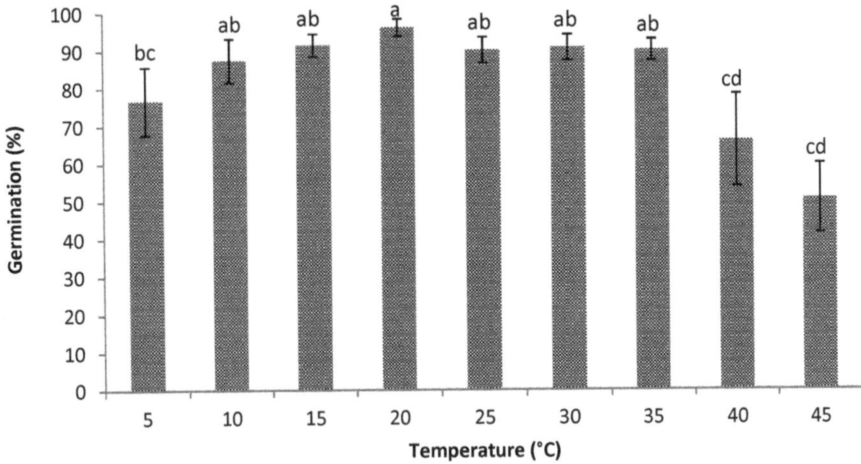

Figure 3. Percentages of germination at each temperature. Error bars show standard deviation. Means that share a letter are not significantly different ($p \leq 0.05$).

The time required for 50% of the seeds to germinate was significantly different among temperature treatments ($F_{8,36}$ 112.34; $p < 0.0001$). To reach 50%, three to five days were required at temperatures above $20 \pm 2\ ^{\circ}\text{C}$, while 18 to 47 days were needed at lower temperatures (15 ± 2 to $5 \pm 2\ ^{\circ}\text{C}$ respectively; Figure 4).

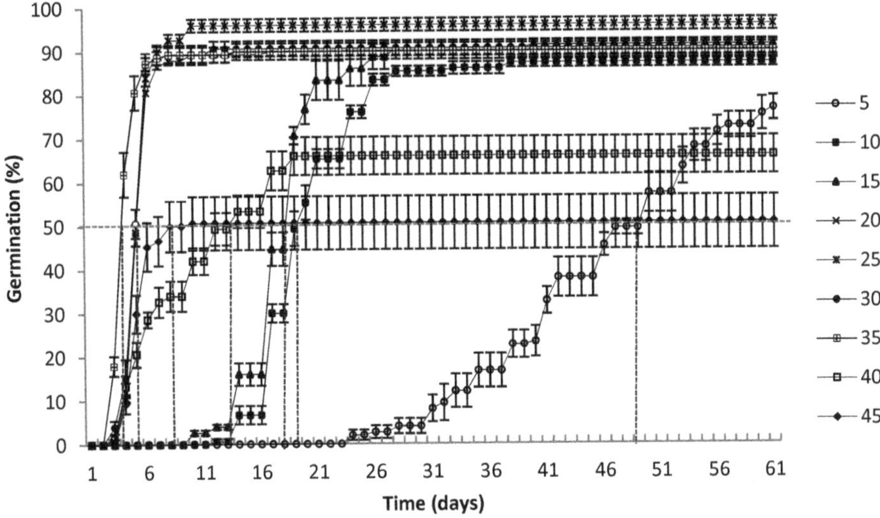

Figure 4. Time required for 50% germination (t_{50}) at each temperature. ($5\ ^{\circ}\text{C}$, $47.41 \pm 3.1\ t_{50}$; $10\ ^{\circ}\text{C}$, $19.76 \pm 2.8\ t_{50}$; $15\ ^{\circ}\text{C}$, $17.86 \pm 2.2\ t_{50}$; $20\ ^{\circ}\text{C}$, $5.02 \pm 0.9\ t_{50}$; $25\ ^{\circ}\text{C}$, $5.02 \pm 0.7\ t_{50}$; $30\ ^{\circ}\text{C}$, $4.83 \pm 0.8\ t_{50}$; $35\ ^{\circ}\text{C}$, $3.76 \pm 1.2\ t_{50}$; $40\ ^{\circ}\text{C}$, $5.34 \pm 2.5\ t_{50}$; $45\ ^{\circ}\text{C}$, $5.65 \pm 0.9\ t_{50}$.). Error bars show standard deviation.

The germination rate differed significantly among temperatures ($F_{8,36} = 28.13$; $p < 0.001$). Germination was fastest at $35\ ^{\circ}\text{C}$, where six seeds germinated per day, and the slowest was at $5\ ^{\circ}\text{C}$, with one seed every two days (Figure 5).

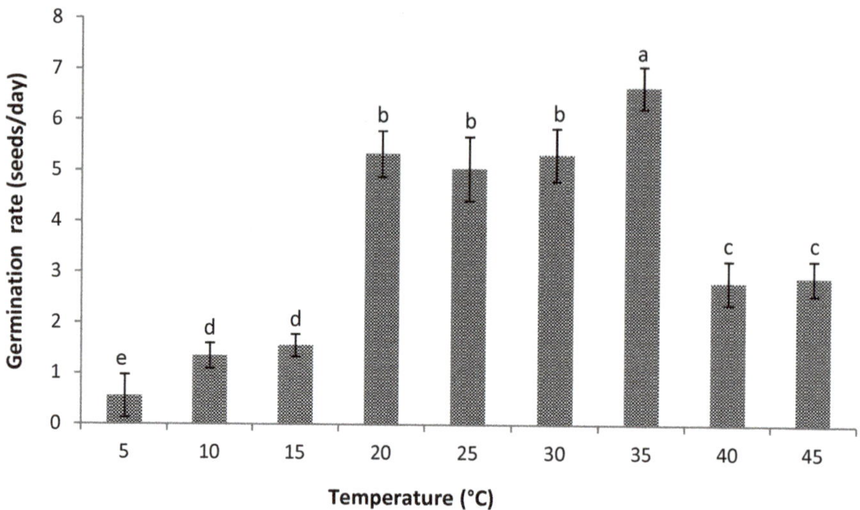

Figure 5. Germination rate per day for each temperature treatment. Error bars show standard deviation. Means that share a letter are not significantly different ($p \leq 0.05$).

3.2. Cardinal Temperatures

Figure 6 shows the germination rates for each temperature. For the sub-optimal temperatures, the model explained 86% of the variation in the germination rate, while for the supra-optimal temperatures the model explained 89% of the variation. With increasing temperature, the germination rate increased until reaching its optimal temperature (To) calculated according to the Hardegree cardinal temperature model [32] of 38 ± 1.6 °C, while decreasing the temperature, decreased the germination rate to zero at the base temperature calculated (Tb) of −0.5 ± 0.09 °C. As temperature increased above the optimal temperature (To), the germination rate decreased to its minimum at the upper threshold (Tc) temperature calculated at 53.3 ± 2.1 °C.

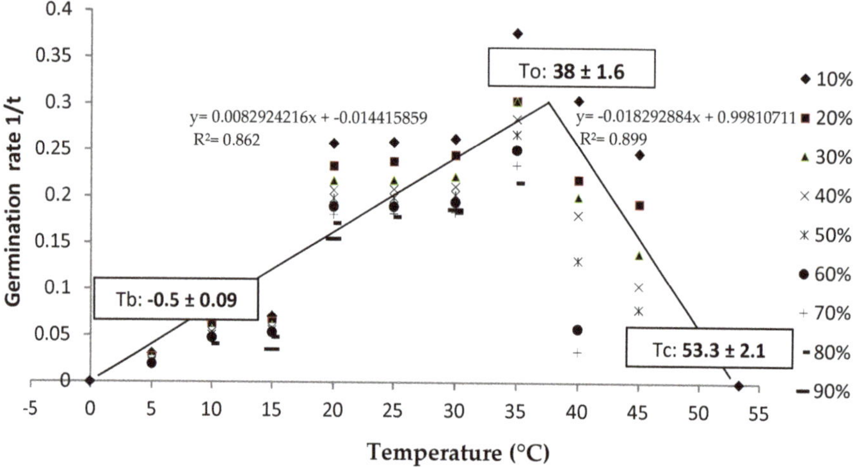

Figure 6. Germination rate in percentiles for each temperature in the range of cardinal temperatures (Tb: Base temperature; To: Optimal temperature; Tc: Upper threshold temperature).

The base temperature was negative, given that more than 75% germination occurred at the lowest temperature tested, 5 ± 2 °C (Figure 3). The optimal temperature (38 ± 1.6 °C) was directly related to the temperature that had the highest germination rate (Figure 5). The upper threshold temperature was above 53 °C, and can be explained due to the fact that at 45 ± 2 °C, it took 5.6 days to germinate 50% of the seed lot (Figure 4). Although the germination percentages are similar in Tb, To and Tc, the significant differences between these calculated values can be observed in the germination rates as shown in Figure 5.

3.3. Thermal Time

According to the base temperature (Tb), in order to obtain 50% of seed germination in the lot, 132.74 ± 2.60 °Cd of thermal time must be accumulated ($\theta_1(50)$).. The probit model explained more than 96% of it. The intercept constant was −4.91 ± 0.53, when the thermal time was zero, and the standard deviation of thermal time was 0.125 ± 0.019 (Table 1). Using the upper threshold temperature (Tc) to obtain 50% of seed germination in the lot, 253.31 ± 4.73 °Cd of thermal time must be accumulated ($\theta_2(50)$), but the probit model explained only 82% of this.

Table 1. Estimated thermal time in seeds of *Cedrela odorata* (Tb = −0.5 °C) from probit regressions in the sub-optimal and supra-optimal temperatures range. θ_1 (50) and θ_2 (50) are shown in log and normal scale.

Parameters	Sub-Optimal	Supra-Optimal
R2	96.33	82.99
K	−4.91 ± 0.53	−2.02 ± 0.28
σ	0.125 ± 0.019	0.122 ± 0.012
$\log\theta$ (50)	2.12 ± 0.008	2.40 ± 0.001
θ (50)	132.74 ± 2.60	253.31 ± 4.73

The valor represent mean ± standard deviation.

For the base temperature (Tb) in Figure 7a, it is evident that the increase in heat accumulation increases the probability of higher germination percentages because it approaches the thermal time ($\theta_1(50)$) of 132.74 ± 2.6, where 50% of the seed lot germinates. In the case of thermal time of the upper threshold temperature (Tc), it indicates that as more heat accumulates, there is an increase in the probability of lower germination percentages. This can be interpreted as movement away from the thermal time ($\theta_2(50)$) of 253.31 ± 4.73 °Cd (Figure 7b).

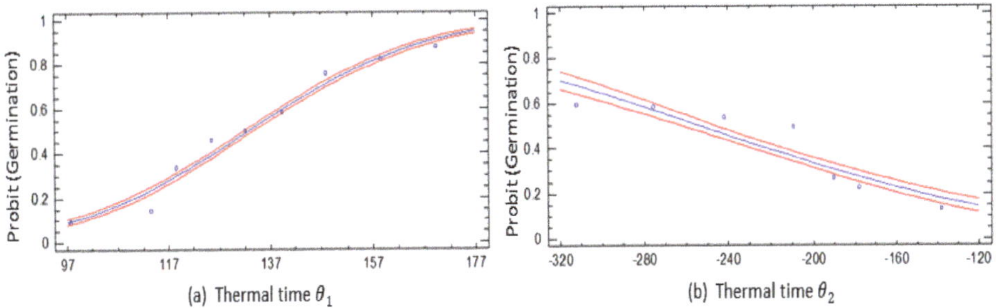

Figure 7. Germination as a function of thermal time for: (**a**) sub-optimal range and (**b**) supra-optimal range of temperatures. The red lines are confidence intervals of germination. The blue line is the estimated data. The points are experimental data.

3.4. Climate Change Scenarios

The current mean temperature for the month of April is 24.5 °C; for the same month in the year 2050, in the German model (MPIECHAM5) the A2 scenario projects a 2 °C and the B2 scenario projects a 1.3 °C increase in temperature. Under the English model (UKMOHADGEM1), scenario A2 projects a 3.1 °C and the B2 scenario projects a 2 °C increase in temperature.

Figure 8 shows the days of the thermal sum of the month of April (the month in which the dispersal of mature seeds begins), based on mean temperature and the increases in temperature under the different climate change scenarios. Under the current scenario, the thermal time is accumulated in 5.5 days. When the temperature increases by 2 °C based on the A2 scenario of the German model (MPIECHAM5), the thermal time accumulates 10 hours earlier than in the current climate. Under the B2 scenario, which projects a 1.3 °C temperature increase, the thermal time accumulates 6.5 hours earlier than the current scenario.

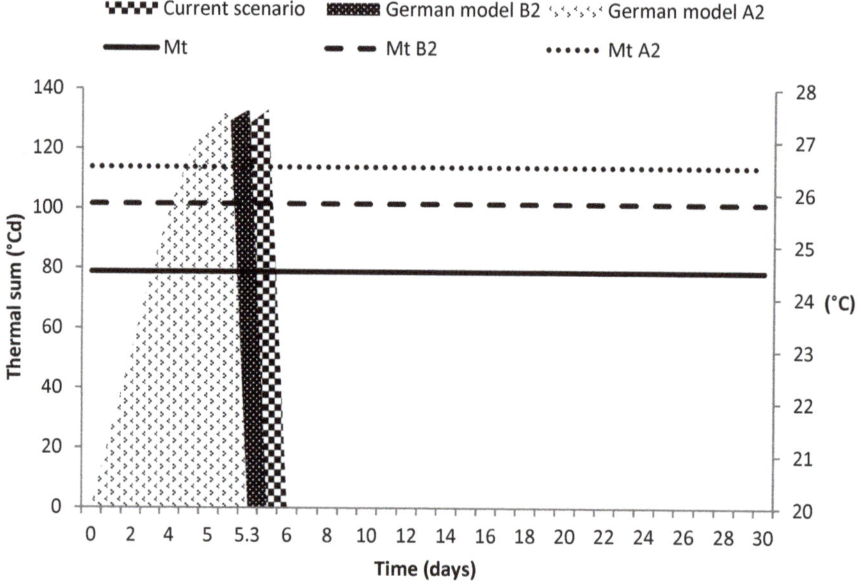

Figure 8. Time in which seeds accumulate the thermal sum (°Cd) during the month of April, for the current climate, and the year 2050 projections under the A2 and B2 scenarios of the German model. (Mt: Current mean temperature; Mt B2: Mean temperature under scenario B2; Mt A2: Mean temperature under scenario A2).

Figure 9 shows results for the English model (UKMOHADGEM1). In scenario A2, which projects a 3.1 °C increase in temperature, the accumulation of the thermal sum is accelerated, so that reaching the thermal time requires just over 4.5 days to germinate 50% of the bank of seeds scattered in the understory. This is 21.5 hours earlier than in the current scenario. For model B2, the thermal time is accumulated 10 hours earlier than in the current scenario.

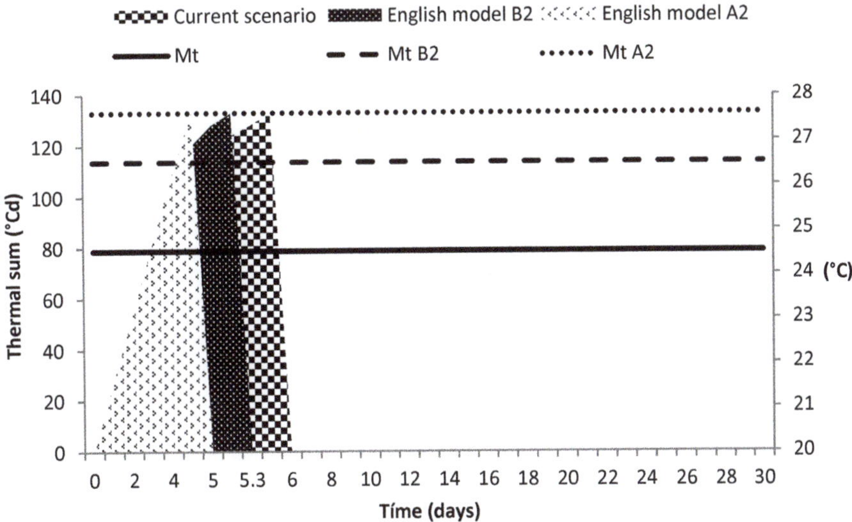

Figure 9. Time in which seeds reach the thermal sum (°Cd) in the month of April, under the current climate scenario and under the 2050 projections for the A2 and B2 scenarios of the English model. (Mt: Current mean temperature; Mt B2: Mean temperature under the B2 scenario; Mt A2: Mean temperature under the A2 scenario).

4. Discussion

4.1. Germination

The highest germination percentage occurred below the theoretical optimal temperature calculated for this species, and does not necessarily occur at the optimal temperature [20]. The highest germination percentage at 20 °C coincides with reports of González et al. [43]. At the same time, the lowest germination percentage occurred in seeds exposed to the highest temperature (45 ± 2 °C), and was significantly lower than that of the lowest temperature tested in this study (5 ± 2 °C). This occurs because, according to Butler et al. [44] very high temperatures (>40 °C) can affect the metabolic processes of the seeds and even damage them, halting embryonic growth, which can inactivate and kill them.

At temperatures at or above 20 ± 2 °C, median germination time was three to five days, which is similar to the results shown by González et al. [43] in *C. odorata* L., and by Manjul and Metali [45] in *Acacia mangium* Wild, *Melastoma beccarianum* Cogn and *Melastoma malabathricum* L. According to Rajjou et al. [46], higher temperature accelerates the enzyme activity in biochemical reactions within the seeds following imbibition. This is true only to a point, after which a further increase in temperature inhibits germination. For this reason, Durán-Puga et al. [47] mention that shorter median germination times may confer a competitive advantage for colonization of fragmented habitats. In addition, Baskin and Baskin [48], suggest that selective pressure in extreme environments increases establishment success.

Seeds germinated at temperatures below 15 ± 2 °C had lower germination rates, even slower than at the highest temperatures, such as 45 ± 2 °C. This coincides with the 15 °C reported by González et al. [43] for *C. odorata* in neotropical dry forest. This can be explained by the fact that, according to Adam et al. [49], lower temperatures decrease metabolic rates to the point that the processes that are essential to germination decrease. Furthermore, this species needs a higher temperature for its development due to its distribution in tropical climates. In addition, Caroca et al. [50] mention that at high temperatures, chemical reactions are faster because water is absorbed more quickly into the seeds, causing the seeds to germinate faster.

4.2. Cardinal Temperatures

C. odorata presented a wide range of cardinal temperatures, from −0.5 °C to above 53 °C. According to Lindig-Cisneros [51], terrestrial vascular plants can tolerate a temperature range in terms of survival from −5 to 60 °C. As such, the ability to germinate at a wide range of temperatures is an adaptive strategy of the species, which gives it an advantage in terms of its potential distribution under different climate change scenarios. According to Parmoon et al. [31], with increasing temperature, there is a linear increase in germination percentage up to the optimal temperature. If temperature increases further, the germination will decrease to zero.

The optimal temperature was 38 ± 1.6 °C. This temperature is not registered in the study area, even in the warmest month, so these results suggest the physiological optimum in lab conditions, rather the ecological optimum temperature [52]. These findings are consistent with Caroca et al. [50], who showed that optimal temperature occurs at higher germination rate. In addition, Adam et al. [49] and Grey et al. [53] indicate that the germination rate increases with temperature, but when the temperatures are above the optimal temperature there is a decrease in the germination rate, specifically at the intersection where the optimal germination temperature is found.

The upper threshold temperature was above 53 °C, which could constitute a competitive advantage for the species when competing to colonize fragmented habitats, since according to Sánchez-Rendón et al. [54] the soil in forest clearings can reach temperatures above 45 °C. In addition, Cóbar-Carranza et al. [55] report germination percentages above 90% in *Pinus contorta* at temperatures between 60 °C and 80 °C.

The base temperature was −0.5 °C. This could be explained with a germination percentage above 75% at 5 ± 2 °C. According to Mendizábal-Hernández et al. [2], this species has a wide distribution due to its plasticity and adaptation to different environmental conditions. According to Calzada-López et al. [20], after imbibition, low temperatures decrease seed metabolism activity and protein synthesis, thus more days are needed to accumulate heat for the germination process. Parra-Coronado et al. [56] mention that below the base temperature, phenological development and metabolic processes stop. As such, according to Andreucci et al. [57] the base temperature predicts precisely the dates at which different stages in the phenological development occur in a species.

4.3. Thermal Time

The thermal time required for 50% germination of *C. odorata* seeds was 132.74 ± 2.60 °Cd. This is consistent with findings obtained by Normand and Léchaudel [58] and Parra-Coronado et al. [56] in some tropical species, such as *Psidium* spp. and *Manguifera* spp., which require less time to reach the different phenological stages. According to Funes et al. [59], germination is accelerated by high temperatures in tropical climates, which maximizes individuals' establishment and survival. Furthermore, Asseng et al. [60] mention that in warm climates, phenological development is accelerated, leading to shorter growth periods, which is an adaptive competitive strategy for the species. According to Colauto-Stenzel et al. [61] and Parmoon et al. [31], thermal time is used to include the effect of temperature as the most important bioclimatic factor in regulating the germination process.

4.4. Climate Change Scenarios

Sánchez-Rendón et al. [54] and IPCC [62] consider that increasing temperature and decreasing precipitation will give rise to evolutionary changes and altitudinal and latitudinal migration among forest species. The A2 scenarios of the German and English models project a larger increase in temperature due to high emissions of greenhouse gases, which shortens the necessary time needed to accumulate the thermal sum (°Cd). This is consistent with results of Rajjou et al. [46], who indicate that at higher temperatures, the germination rate increases. According to Funes et al. [59], faster germination assures the establishment of individuals that must compete for space. However, according to Grey et al. [53], very high temperatures decrease germination because they damage the embryo.

The climate change scenarios, according to Sánchez-Rendón et al. [54], in the medium term will present increased temperature and decreased precipitation. However, they also predict extreme events such as droughts and floods, which will impact the function of tropical ecosystems, affecting flowering, fruiting, germination, and establishment periods of forest species. With the increase in temperature, it is expected that the dispersion of *C. odorata* seeds will be in a shorter period, likely with shorter time spans for the accumulation of thermal time necessary for the phenological stages of the species.

Gutiérrez and Trejo [35] state that climate change will alter the distribution and abundance of species because they will be forced to migrate, altering the sustainability of the region. This is consistent with projections for *C. odorata*; according to Romo-Lozano et al. [3] the timber stocks of this species in Mexico will be valued at $5,282,183,403 Mexican Pesos, and 68% of the stock will be found in the Gulf of Mexico region. However, according to the results of the present work, these projections need to be reanalyzed, since the wide range between Tb and Tc temperatures implies significant changes in their distribution, and in the potential forest areas. In addition, Hernández-Ramos et al. [63], predict that by the year 2050 a reduction in the current area reported for *C. odorata* in Tabasco and the Yucatan Peninsula, as well as the southern part of Veracruz and Chiapas, will occur. However, there is a high probability that it will be distributed in the southern part of the Yucatan Peninsula, northern Chiapas and the coastal plain of the Gulf of Mexico in Veracruz. Furthermore, Gómez-Díaz et al. [11] mention that they expect a reduction in the species' distribution in the state of Hidalgo, due to the increase in temperature and decrease in precipitation. However, the species is also expected to adapt to the new limited precipitation conditions with possible altitudinal and latitudinal migration [25].

To our knowledge, this is the first study of cardinal temperatures and thermal time in seeds of *C. odorata*, which is one of the most important forest species in tropical environments, since its wood is highly valuable in carpentry and cabinetmaking. Additionally, the study will be the basis for the generation of bioclimatic models that can be used to predict the potential distribution of the species, under extreme ecological conditions, in order to develop conservation strategies for fragile ecosystems, and restore fragmented habitats.

5. Conclusions

C. odorata presented a wide range of cardinal temperatures; the base temperature was half a degree below zero and the maximum threshold temperature of 53.3 °C. while the optimum temperature for germination was 38 °C. With results obtained, a thermal time (θ_1 (50)) of 132.74 °Cd could be calculated for the germination stage, which in the current scenario accumulates in 5.4 days. Thus, as the temperature increases according to the climate change scenarios MS-2050 and MS-2100, the germination of the soil seed bank in the undergrowth will accelerate. Because of these findings, it is advisable to evaluate the cardinal temperatures in this work in successive developmental stages such as establishment and growth to understand predicted reduction in its distribution.

Author Contributions: Conceptualization: S.S-M., C.A.O.-S., P.D.D.-A., R.L.-S., O.T.-V., and C.M.F.-O. Data curation: N.I.R.-A. Formal analysis: S.S.-M., C.A.O.-S., R.L.-S., O.T.-V., and C.M.F.-O. Funding acquisition: T.U., M.W., and P.D.D.-A. Investigation: S.S.-M., T.U., and C.M.F.-O. Methodology: C.A.O.-S., and N.I.R.-A. Project administration: T.U., P.D.D.-A., and N.I.R.-A. Resources: M.W., and N.I.R.-A. Writing—original draft: S.S.M., and C.M.F.-O. Writing—review & editing: C.A.O.-S., E.M., T.U., M.W., E.C.-L., P.D.D.-A., R.L.-S., and O.T.-V.

Funding: This study has been funded by the Garfield Weston Foundation, as part of the Global Tree Seed Bank Project, managed by the Royal Botanic Gardens, Kew.

Acknowledgments: We thank the Postdoctoral Fellowship Program at the UNAM, of the General Direction of Academic Personnel Affairs. This study has been funded by the Garfield Weston Foundation, as part of the Global Tree Seed Bank Project, managed by the Royal Botanic Gardens, Kew.

Conflicts of Interest: The authors declare that they have no conflict of interest.

References

1. Pennington, T.D.; Sarukhán, J. *Tropical Trees of Mexico: Manual for the Identification of the Main Species*, 3rd ed.; University Scientific Text. UNAM: Mexico City, Mexico, 2005; pp. 294–296.
2. Mendizábal-Hernández, L.C.; Alba-Landa, J.; Márquez-Ramírez, J.; Cruz-Jiménez, H.; Ramírez-García, E.O. Carbon sequestration by *Cedrela odorata* L. in a genetic trial. *Revista Mexicana de Ciencias Forestales* **2011**, *2*, 105–111.
3. Romo-Lozano, J.L.; Vargas-Hernández, J.J.; López-Upton, J.; Ávila-Angulo, M.L. Estimate of the financial value of timber stocks of red cedar (*Cedrela odorata* L.) in Mexico. *Madera y Bosques* **2017**, *23*, 111–120. [CrossRef]
4. Navarro, C.; Montagnini, F.; Hernández, G. Genetic variability of *Cedrela odorata* Linnaeus: Results of early performance of provenances and families from Mesoamerica grown in association with coffee. *For. Ecol. Manag.* **2004**, *192*, 217–227. [CrossRef]
5. Ramírez-Garcí, C.; Vera-Castillo, G.; Carrillo-Anzures, F.; Magaña-Torres, O.S. Red cedar (*Cedrela odorata* L.) as reconversion alternative of agricultural lands in the south of Tamaulipas. *Agricultura Técnica en México* **2008**, *34*, 243–256.
6. Pérez-Salicrup, D.R.; Esquivel, R. Tree infection by *Hypsipyla grandella* in *Swietenia macrophylla* and *Cedrela odorata* (*Meliaceae*) in Mexico's southern Yucatan Peninsula. *For. Ecol. Manag.* **2008**, *225*, 324–327. [CrossRef]
7. Sampayo-Maldonado, S.; Jiménez-Casas, M.; López-Upton, J.; Sánchez-Monsalvo, V.; Jasso-Mata, J.; Equihua-Martínez, A.; Castillo-Martínez, C.R. *Cedrela odorata* L. mini-cutting rooting. *Agrociencias* **2016**, *50*, 919–929.
8. Comisión Nacional Forestal (CONAFOR). Main Timber Species Established in Commercial Forest Plantations, 2000–2014. Available online: http://www.conafor.gob.mx:8080/documentos/ver.aspx?grupo=43&articulo=6022 (accessed on 15 July 2018).
9. Convention on International Trade in Endangered Species of Wild Fauna and Flora (CITES). In Proceedings of the Fourteenth meeting of the Conference of the Parties, The Hague, The Netherlands, 3–15 June 2007; p. 26.
10. Secretaria del Medio Ambiente y Recursos Naturales (SEMARNAT). Norma Oficial Mexicana NOM-059-SEMARNAT-2010, Protección Ambiental-Especies Nativas de México de flora y Fauna Silvestres-Categorías de Riesgo y Especificaciones para su Inclusión, Exclusión o Cambio-Lista de Especies en Riesgo. Available online: https://www.gob.mx/cms/uploads/attachment/file/134778/35._NORMA_OFICIAL_MEXICANA_NOM-059-SEMARNAT-2010.pdf (accessed on 25 July 2018).
11. Gómez-Díaz, J.D.; Monterroso-Rivas, A.I.; Tinoco-Rueda, J.A. Distribution of red cedar (*Cedrela odorata* L.) in the Hidalgo state, under current conditions and scenarios of climate change. *Madera y Bosques* **2007**, *13*, 29–49. [CrossRef]
12. Hartmann, H.T.; Kester, D.E.; Davies, F.T.; Geneve, R.L. *Plant Propagation: Principles and Practices*, 8th ed.; Prentice-Hall: Upper Saddle River, NJ, USA, 2013; 913p.
13. González, J.E. Collection and germination of seeds from 26 species of tree in a tropical moist forest. *J. Biol. Trop.* **1991**, *39*, 47–51. [CrossRef]
14. Dewan, S.; Vander-Mijnsbrugge, K.; De Frene, P.; Steenackers, M.; Michiels, B.; Verheyen, K. Maternal temperature during seed maturation affects seed germination and timing of bud set in seedlings of European black poplar. *For. Ecol. Manag.* **2008**, *410*, 126–135. [CrossRef]
15. Quinto, L.; Martínez-Hernández, P.A.; Pimentel-Bribiesca, L.; Rodríguez-Trejo, D.A. Alternatives to improve seed germination in three tropical trees. *Revista Chapingo Serie Ciencias Forestales y del Ambiente* **2009**, *15*, 23–28.
16. Garcia-Huidobro, J.; Monteith, J.L.; Squire, G.R. Time, temperature and germination of pearl millet (*Pennisetum typhoides* S.&H.). I. Constant temperature. *J. Exp. Bot.* **1982**, *33*, 288–296. [CrossRef]
17. Ellis, R.H.; Covell, S.; Roberts, E.H.; Summerfield, R.J. The influence of temperature on seed germination rate in grain legumes. II. Intraspecific variation in chickpea (*Cicer arietinum* L.) at constant temperatures. *J. Exp. Bot.* **1986**, *37*, 1503–1515. [CrossRef]
18. Bradford, K.J. Applications of hydrothermal time to quantifying and modeling seed germination and dormancy. *Weed Sci.* **2002**, *50*, 248–260. [CrossRef]

19. Nakao, E.A.; Cardoso, V.J.M. Analysis of thermal dependence on the germination of braquiarão seeds using the thermal time model. *Braz. J. Biol.* **2016**, *79*, 162–168. [CrossRef] [PubMed]

20. Calzada-López, S.G.; Kohashi-Shibata, J.; Uscanga-Mortera, E.; García-Esteva, A.; Yáñez-Jiménez, P. Cardinal temperatures and germination rate in husk tomato cultivars. *Revista Mexicana de Ciencias Agrícolas* **2014**, *8*, 1451–1458.

21. Ruíz-Corral, J.A.; Flores-López, H.E.; Ramírez-Díaz, J.L.; González-Eguiarte, D.R. Cardinal temperatures and length of maturation cycle of maize hybrid H-311 under rainfed conditions. *Agrociencia* **2002**, *36*, 569–577.

22. Muellner, A.N.; Pennington, T.D.; Chase, M.W. Molecular phylogenetics of Neotropical *Cedrela mahogany* (family, *Meliaceae*) based on nuclear and plastid DNA sequences reveal multiple origins of "*Cedrela odorat*". *Mol. Phylogenet. Evol.* **2009**, *52*, 461–469. [CrossRef] [PubMed]

23. Yang, Q.H.; Wei, X.; Zeng, X.L.; Ye, W.H.; Yin, X.J.; Zhang-Ming, W.; Jiang, Y.S. Seed biology and germination ecophysiology of *Camellia nitidissima. For. Ecol. Manag.* **2008**, *255*, 113–118. [CrossRef]

24. Sánchez-Monsalvo, V.; Salazar-García, J.G.; Vargas-Hernández, J.J.; López-Upton, J.; Jasso-Mata, J. Genetic parameters and response to selection for growth traits in *Cedrela odorata* L. *Fitotecnia Mexicana* **2003**, *26*, 19–27.

25. Castellanos-Acuña, D.; Vance-Borland, K.W.; St. Clair, J.B.; Hamann, A.; López-Upton, J.; Gómez-Pimeda, E.; Ortega-Rodríguez, J.M.; Sáenz-Romero, C. Climate-based seed zones for Mexico: Guiding reforestation under observed and projected climate change. *New For.* **2008**, *49*, 297–309. [CrossRef]

26. Espitia-Camacho, M.; Araméndiz-Tatis, H.; Cardona-Ayala, C. Viability, morphometric, and anatomical characteristics of *Cedrela odorata* L. and *Cariniana pyriformis* Miers seeds. *Agron. Mesoam.* **2017**, *28*, 605–617. [CrossRef]

27. García, E. *Modifications to the Climate Classification System of Köppen*, 5th ed.; series #6; Instituto de Geografía, Universidad Nacional Autónoma de México: Mexico City, Mexico, 2004; pp. 19–49.

28. Villarreal-Manzano, L.A.; Herrera-Cabrera, B.E. Water requirement in the vanilla (*Vanilla planifolia* Jacks. ex Andrews)—Naranjo (*Citrus sinensis* L.) production system in the Totonacapan region, Veracruz, Mexico. *Agroproductividad* **2018**, *11*, 29–36.

29. Sampayo-Maldonado, S.; Castillo-Martínez, C.R.; Jiménez-Casas, M.; Sánchez-Monsalvo, V.; Jasso-Mata, J.; López-Upton, J. In vitro germination of *Cedrela odorata* L. seed from extinct genotypes. *Agroproductividad* **2017**, *10*, 53–58.

30. International Seed Testing Association (ISTA). *International Rules for Seed Testing*; International Seed Testing Association: Zurich, Switzerland, 2005; 243p.

31. Parmoon, G.; Moosavi, S.A.; Akbari, H.; Ebadi, A. Quantifying cardinal temperatures and thermal time required for germination of *Silybum marianum* seed. *Crop J.* **2015**, *3*, 145–151. [CrossRef]

32. Ordoñez-Salanueva, C.A.; Seal, C.E.; Pritchard, H.W.; Orozco-Segovia, A.; Canales-Martínez, M.; Flores-Ortíz, C.M. Cardinal temperatures and thermal time in *Polaskia* Beckeb (Cactaceae) species: Effect of projected soil temperature increase and nurse interaction on germination timing. *J. Arid Environ.* **2015**, *115*, 73–80. [CrossRef]

33. Hardegree, S.P. Predicting germination response to temperature. I. Cardinal temperature models and subpopulation-specific regression. *Ann. Bot.* **2006**, *97*, 1115–1125. [CrossRef] [PubMed]

34. Covell, S.; Ellis, R.H.; Roberts, E.H.; Summerfield, R.J. The influence of temperature on seed germination rate in grain legumes. 1: A comparison of chickpea, lentil, soybean and cowpea at constant temperatures. *J. Exp. Bot.* **1986**, *37*, 705–715. [CrossRef]

35. Gutiérrez, E.; Trejo, I. Effect of climatic change on the potential distribution of five species of temperate forest trees in Mexico. *Revista Mexicana de Biodiversidad* **2014**, *85*, 179–188. [CrossRef]

36. Fernández-Eguiarte, A.; Zavala-Hidalgo, J.; Romero-Centeno, R.; Digital Climatic Atlas of the Mexico. Center of Sciences of the Atmosphere, Universidad Nacional Autónoma de México, México. 2010. Available online: http://uniatmos.atmosfera.unam.mx/ACDM/servmapas (accessed on 13 September 2018).

37. Flores-Magdaleno, H.; Flores-Gallardo, H.; Ojeda-Bustamante, W. Phenological prediction of potato crop by means of thermal time. *Rev. Fitotec. Mex.* **2014**, *37*, 149–157.

38. Orrù, M.; Mattana, E.; Pritchard, H.W.; Bacchetta, G. Thermal thresholds as predictors of seed dormancy release and germination timing: Altitude-related risks from climate warming for the wild grapevine *Vitis vinifera* subsp. *Sylvestris. Ann. Bot.* **2012**, *110*, 1651–1660. [CrossRef] [PubMed]

39. Itoh, A.; Yamakura, T.; Kanzaki, M.; Ohkubo, T.; Palmiotto, P.A.; LaFrankie, J.V.; Kendawang, J.J.; Lee, H.S. Rooting ability of cuttings relates to phylogeny, habitat preference and growth characteristics of tropical rainforest trees. *For. Ecol. Manag.* **2002**, *168*, 275–287. [CrossRef]

40. Syros, T.; Yupsanis, T.; Zafiriadis, H.; Economou, A. Activity and isoforms of peroxidases, lignin and anatomy, during adventitious rooting in cuttings of *Ebenus cretica* L. *J. Plant Physiol.* **2004**, *161*, 69–77. [CrossRef] [PubMed]

41. Muñoz-Gutiérrez, L.; Vargas-Hernández, J.J.; López-Upton, J.; Soto-Hernández, M. Effect of cutting age and substrate temperature on rooting of *Taxus globosa*. *New For.* **2009**, *38*, 187–196. [CrossRef]

42. Statistical Analysis System (SAS). *Institute SAS/STAT 9.1 Guide for Personal Computers*; SAS Institute Inc.: Cary, NC, USA, 2004; 378p.

43. González-Rivas, B.; Tigabu, M.; Castro-Marín, G.; Odén, P.C. Seed germination and seedling stablishment of Neotropical dry forest species in response to temperature and light conditions. *J. For. Res.* **2009**, *20*, 99–104. [CrossRef]

44. Buttler, T.J.; Celen, A.E.; Webb, S.L.; Krstic, D.; Interrante, S.M. Temperature affects the germination of forage legume seeds. *Crop Science* **2014**, *54*, 2846–2853. [CrossRef]

45. Manjul, N.M.J.; Metali, F. Germination and growth of selected tropical pioneers in Brunei Darussalam: Effects of temperature and seed size. *Res. J. Seed Sci.* **2016**, *9*, 48–53. [CrossRef]

46. Rajjou, L.; Duval, M.; Gallardo, K.; Catusse, J.; Bally, J.; Job, C.; Job, D. Seed germination and vigour. *Annu. Rev. Plant Biol.* **2012**, *63*, 507–533. [CrossRef]

47. Duran-Puga, N.; Ruíz-Corral, J.A.; González-Eguiarte, D.R.; Núñez-Hernández, G.; Padilla-Ramírez, F.J.; Contreras-Rodríguez, S.H. Devolopment cardinal temperatures of the planting-emergence stage for 11 forage grasses. *Rev. Mex. Cienc. Pecu.* **2011**, *2*, 347–357.

48. Baskin, C.C.; Baskin, J.M. Seed dormancy in trees of climax tropical vegetation types. *Trop. Ecol.* **2005**, *46*, 17–28.

49. Adam, N.R.; Dierig, D.A.; Coffelt, T.A.; Wintermeyer, M.J.; Mackey, B.E.; Wall, G.W. Cardinal temperatures for germination and early growth of two *Lesquerella* species. *Ind. Crops Prod.* **2007**, *25*, 24–33. [CrossRef]

50. Caroca, R.; Zapata, N.; Vargas, M. Temperature effect on the germination of four peanut genotypes (*Arachis hypogaea* L.). *Chil. J. Agric. Anim. Sci.* **2016**, *32*, 94–101. [CrossRef]

51. Lindig-Cisneros, R. *Ecology of Restoration and Environmental Restoration*; Escuela Nacional de Estudios Superiores, Unidad Morelia. Instituto de Investigaciones en Ecosistemas y Sustentabilidad, Universidad Nacional Autónoma de México: Mexico, 2017; pp. 59–60.

52. Orozco-Segovia, A.; González-Zertuche, L.; Mendoza, A.; Orozco, S. A mathematical model that uses gaussian distribution to analyze the germination of *Manfreda brachystachya* (*Agavaceae*) in a thermogradient. *Physiol. Plant.* **1996**, *98*, 431–438. [CrossRef]

53. Grey, T.L.; Beasley, J.P.; Webster, T.M.; Chen, C.Y. Peanut seed vigor evaluation using a thermal gradient. *Int. J. Agron.* **2011**, *7*. [CrossRef]

54. Sánchez-Rendón, J.A.; Suárez-Rodríguez, A.G.; Montejo-Valdés, L.; Muñoz-Garcia, C. Climate change and the seeds from the Cuban native plants. *Acta Botánica Cubana* **2011**, *214*, 38–50.

55. Cóbar-Carranza, A.J.; García, R.A.; Pauchard, A.; Peña, E. Effects of high temperatures in germination and seed survival of the invasive species *Pinus contorta* and two native species of South Chile. *Bosque* **2015**, *36*, 53–60. [CrossRef]

56. Parra-Coronado, A.; Fischer, G.; Chaves-Cordoba, B. Thermal time for reproductive phenological stages of pineapple guava (*Acca sellowiana* (O. Berg) Burret). *Acta Biol. Colomb.* **2015**, *20*, 163–173. [CrossRef]

57. Andreucci, M.P.; Moot, D.J.; Blakc, A.D.; Sedcole, R. A comparison of cardinal temperatures estimated by linear and nonlinear models for germination and bulb growth of forage brassicas. *Eur. J. Agron.* **2016**, *81*, 52–63. [CrossRef]

58. Normand, F.; Léchaudel, M. Toward a better interpretation and use of thermal time model. *Acta Hortic.* **2006**, *707*, 159–164. [CrossRef]

59. Funes, G.; Díaz, S.; Venier, P. Temperature as a main factor determining germination in Argentinean dry Chaco species. *Ecol. Austral* **2009**, *19*, 129–138.

60. Asseng, S.; Foster, I.; Turner, N.C. The impact of temperature variability on wheat yields. *Glob. Chang. Biol.* **2011**, *17*, 997–1012. [CrossRef]

61. Colauto-Stenzel, N.M.; Janeiro-Neves, C.S.V.; Jamil-Marur, C.; DosSantos-Scholz, M.B.; Gomes, J.C. Maturation curves and degree-day accumulation for fruits of "Folha Murcha" orange trees. *Sci. Agric.* **2006**, *63*, 219–225. [CrossRef]

62. Intergovernmental Panel of Climate Change (IPCC). *Climate Change 2013: The Physical Science Basis;* Contribution of Working Group I to the Fifth Assessment Report of the Intergovernmental Panel of Climate Change; Stocker, T.F., Qin, D., Plattner, G.K., Tignor, M., Allen, S.K., Boschung, J., Nauels, A., Xia, Y., Bex, V., Midgley, P.M., Eds.; Cambridge University Press: Cambridge, UK; New York, NY, USA, 2013; pp. 22–127.

63. Hernández-Ramos, J.; Reynoso-Santos, R.; Hernández-Ramos, A.; García-Cuevas, X.; Hernández-Máximo, E.; Cob-Uicab, J.V.; Sumano-López, D. Historical, current and future distribution of *Cedrela odorata* in Mexico. *Acta Botánica Mexicana* **2018**, *124*, 117–134. [CrossRef]

MDPI
St. Alban-Anlage 66
4052 Basel
Switzerland
Tel. +41 61 683 77 34
Fax +41 61 302 89 18
www.mdpi.com

Forests Editorial Office
E-mail: forests@mdpi.com
www.mdpi.com/journal/forests